SpringerBriefs on Case Studies of Sustainable Development

Series editors

Cecilia Tortajada, Los Clubes, Atizapan, Mexico
Asit K. Biswas, Third World Centre for Water Manage, Los Clubes, Atizapan, Mexico

W0235391

More information about this series at http://www.springer.com/series/11889

Jean-Paul Close

Editor

AiREAS: Sustainocracy for a Healthy City

The Invisible made Visible Phase 1

Co-Authors: Jean-Paul Close, Nicholas A.S. Hamm, Gerard Hoek, Rene Otjes, Benjamin Rosen, Mary-Ann Schreurs, Alfred Stein, Sandra van der Sterren, Marco van Lochem, Hans Verhoeven, Edwin Weijtmans

Editor
Jean-Paul Close
AiREAS
Eindhoven
The Netherlands

ISSN 2196-7830   ISSN 2196-7849   (electronic)
SpringerBriefs on Case Studies of Sustainable Development
ISBN 978-3-319-26939-9   ISBN 978-3-319-26940-5   (eBook)
DOI 10.1007/978-3-319-26940-5

Library of Congress Control Number: 2015957053

Printed on acid-free paper

This Springer imprint is published by SpringerNature
The registered company is Springer International Publishing AG Switzerland

Contents

About the Co-authors

The list of co-authors of this study is long, multidisciplinary, and functional in relation to the responsibility shared in producing this analysis and its progression in co-creating a "Healthy City" together. Here is the list and the affiliation of each in alphabetical order:

Jean-Paul Close Ideological father of the sustainocratic way of working. Initiator of the STIR foundation and co-founder of AiREAS. Jean-Paul is president of STIR and AiREAS, and overall coordinating editor of this publication.

Dr. Nicholas A.S. Hamm is an assistant professor in the Faculty of Geo-Information Science and Earth Observation (ITC), University of Twente, The Netherlands. His research focuses on spatial data quality and statistical modelling for Earth observation data, with a particular focus on health-related applications such as environmental pollution. Nicholas was part of the group that designed the ILM from the perspective of air pollution modelling for scientific research. He is the main author of Chap. 3 which constitutes the main scientific contribution to this publication.

Dr. Gerard Hoek is affiliated with the IRAS institute of the University of Utrecht and internationally recognized as an expert in research on air pollution and human health. Gerard became involved in AiREAS Eindhoven right from day one and plays a key role in the design of the ILM from the perspective of air pollution exposure and measurements for scientific research.

Ir. René Otjes affiliated with the Energy Research Centre of the Netherlands (ECN). He has 35 years of experience in development and application of monitors for air quality parameters. Rene became involved in AiREAS for the purpose of being one of those to define the technological characteristics of the Airbox in the ILM, which he subsequently developed and put into practice. He also plays a key role in balancing the commercial drive of ECN with the committed responsibility of co-creating healthy city deliverables.

Benjamin Rosen a former teacher of mathematics and science who is currently starting a new career in permaculture design, connected with STIR and AiREAS late in the process. Once acquainted with the ideology and practical contributions we make, he offered to contribute an analysis of the economic processes that led to the situation today. His analysis proved to be a valuable contribution for completing the manuscript and positioning the awareness timeline of initiator Jean-Paul Close.

Mary-Ann Schreurs had just been installed as Alderman of the city of Eindhoven when she was asked to participate in AiREAS in 2010. The challenge fits her personal leadership ambition of guiding Eindhoven into a new reality with citizen participation, sustainable progress, and innovation as key policy instruments. She has subsequently played a key role in transforming the hierarchical positioning of the city executive into a participative partnership, facilitating development of the co-creation platform AiREAS.

Prof. Dr. Alfred Stein is head of the ITC department of the University of Twente and got involved with AiREAS right at the early conceptual stage. He has a worldwide reputation in the field of research on spatial modelling, especially in defining measurement infrastructures for air pollution and its spatial dispersion. His presence on the team was key to determining the early characteristics and design of the ILM.

Sandra van der Sterren Within the Department of the Environment of the city of Eindhoven, Sandra carries the responsibility for air quality. She played a key role in defining the measurement infrastructure of the ILM. Once the network became operational, Sandra became key to the multidisciplinary platform of analysis and interpretation of data and development of innovations and policies.

Marco van Lochem is co-founder of AiREAS. He first became affiliated with the STIR foundation "to give something back to society" and was the first professional to second this "Healthy City" initiative. At the time of the co-founding of AiREAS, he was self-employed within a great network of technology-driven organizations, especially ICT and the Internet of Things. During his work for AiREAS, his personal evolution brought him into executive positions in organizations such as Imtech and later Axians (Vinci). As vice president of AiREAS, he plays a key role in the strategic connection of technological partner hierarchies while also overseeing administrative economic responsibilities.

Hans Verhoeven affiliation is with the city government of Eindhoven for which he is the leader of environmental programs. He was asked by Mary-Ann Schreurs to take on the operational task of representing the city government in this team effort. His role was key to getting the AiREAS projects greenlit, backed by the corresponding government departments and formally followed up in the coalition.

Edwin Weijtmans In his role back in 2010 as program manager for air quality in the province of Noord Brabant, Edwin became the very first government official to commit personally to the co-creation of AiREAS, helping it transform from ideology to proof of concept and aiding in the completion of this first phase. Edwin played a key role at his institution, creating awareness of the AiREAS programs, garnering executive support, and providing part of the finances to make it all happen.

All those who have contributed to this first phase of AiREAS, deployed in Eindhoven, are also playing a central role in expanding the AiREAS principle into other regions of the Netherlands, Europe, and the world.

A short mention must be made of our gratitude to

Erik van Merrienboer was an executive member of the city council of Eindhoven and an early contact of STIR. After his term in the city, he became director of Mobility and Economy in the province of Noord Brabant. He played a significant role in the AiREAS proof of concept (2010) phase when the initiative was still positioned at a national rather than a city level. After the proof of concept, the focus was placed on the city of Eindhoven for phase 1. Erik kept following the process as the province remained committed through the efforts of Edwin Weijtmans.

Introduction

This project started with a mission: A mission to transform ourselves, citizens of a twenty-first-century post-industrial consumer society driven by self-interest, greed, the need for instant gratification, and the desire for flashy ephemeral things, into new men and women, inspired by human values, motivated by common purpose, seeking to build a sustainable world order, shouldering shared responsibilities, unleashing our creative energies, and striving for better ways of being.

In this report, we describe the rather unknown world of the conscience, the way our awareness opens us up to new views of our reality, including our own humble beginnings as a value-driven community. We explain how our mission for the co-creation of "Healthy Cities" eventually came about in a multidisciplinary cooperative process, driven by human values. We describe how the STIR foundation[1] came into being after a process of trial and error, and how, from this foundation, the Global AiREAS project was established. The Local AiREAS project in Eindhoven has become our living laboratory, a source of inspiration and development of unprecedented expertise. Here, we show how the first co-creative phase, which we called "making visible the invisible," unfolded and how this portends a new way of positioning individuals and groups in society.

Then, we describe the practical realization of the first step in Global AiREAS, culminating in the completion of phase 1, the development and deployment of the Innovative Air Quality Measurement System (ILM—Innovatief Lucht Meetsysteem).[2]

Here, we share our experiences. We start with the growth of awareness and commitment to transformative change and proceed through a complex group process to fruition. We wish to position AiREAS in a human value-driven and transformative context, including a survey on our search as to why we, as

---

[1]Sustainocratic Transformation, Indexation and Research (STIR) is a foundation that sets up multidisciplinary cooperations to address the global issues that threaten human communities around the world. STIR was founded by Jean-Paul Close in 2009 in response to the credit crisis.

[2]We use this acronym, ILM, from the Dutch name, *Innovatieve Lucht Meetsysteem*.

individuals or institutions, connect to such ventures, or why not. We even believe that this multidisciplinary commitment, initiated by citizens and citizen groups, seconded by institutions and government officials, is unprecedented. The process is described from the early conception of how individuals and groups could connect emotionally and ideologically to a new human value system for society, up to the moment it was realized through a practical project that encompasses an entire city and draws upon the contributions of multiple partners: private, public, and corporate.

Together, we feel the need to document this progress, at the moment in which we finalize the technological ILM project, as an acknowledgement of the partners involved, and to register our accomplishments in order to aid our fractal expansion into other areas or cities in the world. After this, we can turn to new cycles in this human value-driven enterprise. We intend to use this first phase as a template for further AiREAS steps in the city of Eindhoven, and to implement phase 1 (ILM) in other cities globally. There are many people and organizations involved, directly and indirectly. This document will become an historical reference for AiREAS and other human value-driven initiatives of similar complexity of which the STIR foundation already has many examples.[3]

Thus, this publication contains four parts:

1. *A theoretical review of economic principles relevant to community-based projects that take control of open access resources in the commons*. In this potted review of economic theory, we race through four hundred years of liberal economic thinking to show why the systems we have today are failing both us and the planet, and we point the way to the future.
2. *The ideological and practical realization of the City of Tomorrow initiative and the subsequent foundation of AiREAS*, the "Healthy City" purpose, and working format up to the point of deciding to "make the invisible visible" by designing, developing and implementing the innovative ILM, phase 1, the *Innovative Air Quality Measurement System (ILM)* in the city of Eindhoven.
3. *The scientific and technological choices* made in the process of designing and implementing the ILM system, up to the realization of implementation in the field.
4. *The practical experience* with the system after one year of functioning, its link with the original assumptions and desires, the experiences developed in the process of co-creative, multidisciplinary interaction, the lessons learned, spin-offs, points of unique excellence, and issues for improvement.

---

[3]Other STIR initiatives to date are as follows: FRE2SH (eco-city: local self-sufficiency and productivity), STIR Academy (educational triple "i" platform: inspiration, innovation, implementation), and SAFE (safety and social innovation).

This document has been co-authored by the key people involved in AiREAS and the phase 1 project ILM of Local AiREAS Eindhoven. All credit goes to the multidisciplinary partnership, comprising members that had the courage, authority, and determination to accept the invitation to undertake this co-creation and bring it to its unique result-driven enterprise for healthy eco-cities.

Jean-Paul Close
Co-author and editor

Chapter 1
Potted Review of Economic Theory: The Complex Evolving System

Benjamin Aaron Rosen

1.1 A Potted Review

Humans live in material intercourse with both nature and each other to sustain life. These arrangements, which broadly fall under the notion of 'economics', have not always bankrupted nature. The long hunter-gatherer phase of human existence, in general, left a small ecological footprint. Nevertheless, many civilizations and even smaller-scale societies have destroyed their ecosystems through loss of topsoil, extinction of over-exploited species or failure of overly elaborate and inflexible arrangements; for example, dependence on complex irrigation systems that fail when drought strikes.[1] The rise of fossil-fuel based economies, originating in the West in the 19th century and expanded globally by the late 20th, has produced a rise in living standards and population that is unprecedented, but that rise is now hitting ecological limits. How have Western ideas of economics dovetailed with this economic dynamism and ecological destruction? What can be gained by going back to models that, in terms of their human balance in nature, and perhaps their balanced relations among people as well, were more successful? In this chapter, and in this analysis as a whole, we will address these questions.

The original version of this chapter was revised. The Erratum to this chapter is available at 10.1007/978-3-319-26940-5_5

[1]Diamond, J. (2011) *Collapse: How Societies Choose to Fail or Succeed.* Penguin Books; Revised edition.

B.A. Rosen (✉)
University of Haifa, Haifa, Israel

© The Author(s) 2016
J.-P. Close (ed.), *AiREAS: Sustainocracy for a Healthy City,*
SpringerBriefs on Case Studies of Sustainable Development,
DOI 10.1007/978-3-319-26940-5_1

When Adam Smith,[2] James Mill,[3] David Ricardo[4] and other intellectuals of the 18th and 19th centuries first described the foundations of the free market system, they gave us an account of how the economic machinery that had come into being as a consequence of the Enlightenment and the emerging democratic forms of government constituted around notions of human rights to life, liberty, property and equality, represented the liberation of mankind from the tyranny of kings and priests. In a free market, each individual makes choices by allocating their own resources as they see fit, voting in the perpetual democracy of the market that, in turn, controls the allocation of natural and human resources to best supply demand. The self-interest of individuals is harnessed by market mechanisms to serve the common good. In Adam Smith's words:

> It is not from the benevolence of the butcher, the brewer, or the baker that we expect our dinner, but from their regard to their own self-interest. We address ourselves not to their humanity but to their self-love, and never talk to them of our own necessities, but of their advantages.

Demand is a word used by economists, but it is not a dry monetary term. It is the expression of desire, the condensation of the needs and wants of many individuals. It is a psychological variable. Demand is one half of the collective process that automatically adjusts economic activity throughout the economy. Supply, the collective willingness to provide, is likewise a description of the psychological states of mind of the complementary agents in acts of trade. In a free market, the meeting of supply and demand at agreed-upon prices can be thought of as an expression of liberty and the right to own and trade one's property. The automatic mechanism of the market is an invisible hand, an idiom borrowed from Adam Smith, that directs human affairs to desirable ends.

The Italian economist, Vilfredo Pareto,[5] argued with the help of mathematical models borrowed from engineering that an unencumbered free market leads to the best possible allocation of resources for human needs; that at its optimum, any change of price will lead to lower net wealth creation and less efficient allocation of resources. Changes by means of taxation, price fixing or subsidies cause adjustments throughout the system. There will be losers and winners, but the sum of the losses will be greater than the sum of the gains, as the market is driven away from the clearing prices that would be agreed upon within an unencumbered free market. Today, this is known as a Pareto efficient market.[6]

---

[2]Smith, A. (1776) *An Inquiry into the Nature and Causes of the Wealth of Nations.* Strahan and Cadell, London.

[3]Mill, J. (1813) *Money and Exchange.* Edinburgh Review.

[4]Ricardo, D. (1817) *On the Principles of Political Economy and Taxation.* John Murray, London.

[5]Pareto, V. (1909) *Cours d'Économie Politique*: Nouvelle édition par G.-H. Bousquet et G. Busino, Librairie Droz, Geneva, 1964, pages 299–345.

[6]Mas-Colell, A., Whinston, M.D., Green, J.R. (1995), "Chapter 16: Equilibrium and its Basic Welfare Properties", *Microeconomic Theory.* Oxford University Press, ISBN 0-19-510268-1.

These early promises of liberation and optimisation have not stood the test of time. There are many reasons why free markets do not achieve the optimum results that had once been hoped for. For example, the Pareto efficient economy does not necessarily lead to the most just distribution of goods. Remember, the market is a kind of economic democracy in which we vote by means of the allocation of our resources. This is not an egalitarian democracy; rich individuals have a greater impact on the market equilibria than do poor individuals. If buyers or sellers are monopolies, or co-operating oligopolies, the pricing mechanism does not work equitably. Furthermore, the necessities of life may force us to allocate resources in ways we would rather not. We cannot, for example, stop eating because we find the cost of food too high. The opportunities to exploit workers abound, a problem that is not restricted to the Dickensian era of the British industrial revolution, as the exploitation of workers today by companies like McDonald's and Walmart attest.

Jeremy Bentham's moral imperative, the greatest happiness for the greatest number, implies that great suffering by a minority may be justified if it contributes to greater happiness for a majority. This could be interpreted as a justification for the persecution of minorities. Similarly, Pareto efficient markets are insensitive to the needs of the poor. Indeed, a Pareto efficient market could exist in which the majority remains poor while a minority becomes fabulously wealthy. The pursuit of maximum GDP is not the same as the pursuit of social justice.

Achieving Pareto efficiency, irrespective of whether this leads to a just allocation of resources, presumes that all costs and benefits of economic activity enter into the pricing mechanism. However, this is not so. Costs and benefits that do not enter into the pricing mechanism are referred to as externalities.[7] External costs are often imposed upon the commons or are paid for by individuals who gain no benefit from the goods and services that are produced. When a forest is cut down to make wood for building houses, the people who purchase the houses pay for the costs incurred throughout the production chain. These include the cost of paying for the lumberjacks who cut down the trees, the transport of logs to saw mills, the milling and curing of the wood, the transport of wood to wholesale and retail outlets, the building labour to create the houses, and so on. Other costs, such as cost of energy, enter into the price of the houses by entering into the pricing mechanism at each step at which energy is consumed. However, when the denuded hillsides where forests once stood are no longer available for pleasure, hunting or as habitat for some species, losses are incurred that have not been paid for in the price of the houses. Imagine that the hills become eroded and unstable, as a result of which landslides triggered by heavy rains bury a village in the valley below, and villagers lose their lives and homes. These damages are also prices paid that do not enter into the price of the houses that have been built of the wood harvested from those forests. Likewise, the carbon that was discharged into the atmosphere to supply energy for the production process contributes to climate change and the concomitant costs that this will impose on future generations. These, too, do not enter

---

[7]Pigou, A.C. (1920) *The Economics of Welfare.* Macmillan and Company, London.

into the price of the houses. Since all economic activity in free markets entails externalities, we can no longer speak of Pareto efficiency as being a meaningful guide for optimum allocation of resources. Instead, we must acknowledge that we have created a system that excels in the pillaging of natural resources, the destruction of nature and the exploitation of certain members of our society. The invisible hand is blind, but unlike Justice, it is often arbitrary in whom it punishes and whom it rewards.

There are other ways in which free markets fail. An ideal market is populated by many independent sellers and buyers, each with perfect knowledge of the market, and all acting as rational agents seeking their own best advantage. However, this is not so in the real world. Few people have perfect knowledge of the markets, and therefore usually make less than optimum choices. Some people make irrational choices because of psychological failings, such as addictions and unwholesome habits. In a world that feeds appetites with temptatious advertising, even the most rational among us occasionally make unwise choices and impulse decisions that we later regret. We are too easily misled into habits of self-centred consumerism. Unfortunately, wisdom is not given to all.

Finally, markets fail when the buyers and sellers are no longer concerned with the utility of the commodities they are trading. By utility, economists mean the usefulness of the commodities for their intended purpose. When we purchase a house to live in, we are motivated by the utility of the house. In contrast, when a speculator purchases a house in anticipation of a future rise in market value, the utility of the house plays a secondary role in the decision to trade. Speculative trade leads to market instability. Bubbles form as speculators chase expectations of capital gains, and busts follow when there is a rush to sell in a falling market. This is an intractable problem, for there is no way to detect which trades are utilitarian and which are speculative. It is not possible to legislate against or control speculative trade. A free market, by definition, must be free for all.

1.1.1 Where Has This Taken Us?

Today, we are destroying the natural environment on an unprecedented scale and at a rate exceeding any prior biologically-induced change. In 1989, the world passed a fossil fuel turning point; the quantity of new discoveries of gas, oil and coal deposits became for the first time less than the quantity of these resources consumed. Except for some blips associated with Arctic finds and fracking shale, and a temporary levelling of the curve because of the recent recession, the world's reserves are running down. Our exploitation of fossil fuels has always been unsustainable, but now the end is in sight.[8]

---

[8]IEA (2006) *World Energy Outlook 2006.* Paris and Washington, D.C. Organisation for Economic Co-operation and Development, International Energy Agency.

Contemporary industrial agriculture is unsustainable.[9] It is energy-driven at every phase. Energy is consumed to plough the land, and to apply artificial fertilizer, which is itself largely a petroleum product. More energy is used for seeding. And even more energy is used to apply herbicide, insecticide, and fungicide, and to pump and spray water in irrigated areas. And yet more energy is used to harvest, in processing, packaging and transporting food to market. In many places, agriculture is draining aquifers and salinating the soil. Intense agriculture also destroys soil quality by killing the natural soil ecology, so that each year greater levels of artificial inputs are required to coax a crop from the increasingly dead land. Runoff is despoiling the rivers and oceans with pollutants, and triggering toxic algal blooms at sea. Use of toxic sprays bleeds out into the environment, damaging the natural ecology. Desertification has always followed in the footsteps of mankind, but never more so than today. In summary: Industrial agriculture is consuming large amounts of fossil fuel, putting carbon in the atmosphere, destroying soil quality and draining aquifers, damaging natural ecosystems and despoiling rivers and oceans. A recent UN report[10] concluded that the industrial agricultural sector would be bankrupt today if the full cost of food production were internalized. We are only able to continue with industrial agriculture because a large part of the cost is placed in the commons. Ultimately, these costs are being paid elsewhere, or will have to be paid by future generations.

In the argument above, we briefly addressed the building of wooden houses and industrial agriculture. We must not forget that almost every form of economic activity imposes costs on the commons. Our children and grandchildren will be made to pay some of the costs of our food, housing, education, entertainment, transport and medical care. Indeed, every aspect of our economic lives leaves a trail of debt behind us. In some cases the debt is paid immediately by people in other countries; for example, the export of obsolescent consumer electronic devices to Africa and the Indian subcontinent imposes a burden of pollution in heavy metals and other toxins on far-away people most of us will never see.

What can be done about this? Can we expect morally responsible behaviour to occur spontaneously in any society? Adam Smith noted[11]:

> I have never known much good done by those who affected to trade for the public good.

In 1968, Garrett Hardin described *The Tragedy of the Commons*[12] in a paper on population. Using an example originally devised by William Foster Lloyd,[13] he

---

[9]Hilton, S. (2015) *More Human: Designing a World Where People Come First*. W.H. Allan, London.

[10]Hoffmann, U. et al., (2013) *Wake Up Before it is too Late: Make Agriculture Truly Sustainable Now for Food Security in a Changing Climate*. Trade and Environment Review. United Nations Conference on Trade and Development.

[11]Smith, A. (1776) *An Inquiry into the Nature and Causes of the Wealth of Nations*. Strahan and Cadell, London.

[12]Hardin, G. (1968) *The Tragedy of the Commons*. Science, volume 162, pages 1243–48.

[13]Lloyd, W.F. (1833) *Two Lectures on the Checks to Population*. Oxford University Press, Oxford.

described how a group of farmers competing for use of a shared public resource will over-exploit and under-invest in the commons. Lloyd's model illustrates how it is in the best self-interest of each farmer to extract as much from the commons as possible, even to the point of over-grazing, for by extraction, the individual gains 100 % of the benefit but only a fraction of the losses that are shared by all farmers using the commons. Similarly, it is not in the interest of any one farmer to make a contribution to the commons, since he will pay 100 % of the cost but gain only a fraction of the benefit, while all others sharing the commons will become freeloaders on his contribution. It is therefore in the rational best self-interest of every individual to pillage the resources of the commons without investing in the preservation or development of same.

The Tragedy of the Commons includes the discharge of pollution: chemicals and heat into rivers, toxic waste into the air and oceans, rubbish on the streets and in parks, noise and stink around industrial installations and airports. The polluter gains 100 % of the benefit of being rid of his waste, but suffers only a fraction of the burden. It is therefore in the rational best self-interest of each individual to dispose of his waste into the commons without regard for the price paid by others. And likewise, there is no rational self interest in being the one to do the cleaning up.

The Tragedy of the Commons explains why we inevitably pillage the resources of the earth and foul our planet with waste, each of us in pursuit of our individual self-interest. What is in the best individual self-interest when practiced by many is not in the collective interest to such a degree that individual interests are eventually smothered. As in the prisoner's dilemma, we are driven inexorably to suboptimal outcomes. The logic of the Tragedy of the Commons locks us into a destructive spiral of such vast proportions that we may eventually destroy much of life on this planet.

Solutions to this problem can be placed under two broad headings.

- The Dirigeant Option: Retain the public status of the commons, and allocate permits to exploit the public resources at a level that prevents over-exploitation. Legislate to attach a price to pollution, ranging from light taxes to criminal sanctions.
- The Liberal Option: Privatise the commons, knowing that proprietors are motivated to invest in the development and sustainability of their private property, and generally do not foul their own nests.

Unfortunately, neither of these options satisfactorily solves the problem:

- The dirigeant option presumes that legislative bodies and executive authorities are wise, knowledgeable, incorruptible, attentive and well-intentioned. It also presumes that national authorities have sufficient resources adequately to perform their role as custodians of shared common pool resources. None of these presumptions are true all of the time, often only a few are simultaneously, and in some cases, none are true at all. Whenever you get government attempting to regulate business, you get business attempting to regulate government, corrupting it and undermining democracy. In many countries, the oligarchy of big business becomes the government. This is one of the principal structural problems we have today in most of the world.[14]

- The Liberal option merely pushes the Tragedy of the Commons one step down the road; the newly privatised entities exploit and pollute the remaining commons. Since it is not possible to privatise everything, the problems remain unresolved. Furthermore, this solution plays into the hands of the rich and powerful. The majority remain poor while a minority becomes the inheritors of the riches of the earth.
- In extreme cases, such as in some African countries today, a combination of the dirigeant and liberal options jointly fail. This brings about a collapse of capitalism and a regression to the robber baron phase, a stage of economic development that should have passed into history. The growing wealth-gap in most developed economies also betrays the presence of this disease in these countries.

In cases in which the actions of one entity damage the property of another, these damages being externalities with respect to the activity of the parties inflicting the damage, the usual course of action is to sanction the actions with fines or other forms of legislative control. Ronald Coase[15] argued that where legal rights to open access commons and rights to private property come into conflict, spontaneous local bargaining will occur, leading to internalisation of the costs in the most efficient possible manner. He illustrated this with an example of a cattle rancher negotiating with a crop farmer for access to grazing land. Coase showed that the resolution of disputes that arise when cattle break into farmed land and damage crops will be determined by the relative profitability of cattle-grazing compared to farming. The cattle rancher will agree to pay the farmer for access to grazing land and pay compensation for damages to his crops if the cost is justified by the value of the access and grazing. If, however, the crop is more valuable, then the farmer will demand compensation that the cattle rancher will be unwilling to pay, resulting in a search for other solutions. The cattle rancher might agree to pay for the fence to keep his cattle off the farmer's land if the cost of the fence is less than the compensation that would have to be paid for damaged crops, provided the cost of the fence did not render cattle ranching unprofitable. Coasian bargaining is sensitive to local conditions, is flexible and can adjust to changing circumstances, and incorporates monitoring and sanctions where the participants deem it necessary. Coasian bargaining achieves an economic efficiency that the blunt tool of dirigeant intervention usually cannot.

However, there are three problems with Coasian bargaining. First, the rights of all parties must be established in law before bargaining can take place. Where no statutory rights exist, there are no grounds to force the opposing parties to the bargaining table. We, as people who live in a world increasingly polluted and degraded by the practices of industry, generally have no legal right to seek redress. We suffer in an increasingly polluted world. Second, transaction costs are incurred

---

[14]Barnes, P. (2006) *Capitalism 3.0: A Guide to Reclaiming the Commons*. Berrett-Koehler.

[15]Coase, R.H. (1960) *The Problem of Social Cost*. The Journal of Law and Economics. 3, 1–44.

in the process of bargaining. These are especially high when one party comprises many individuals with small marginal interests in the outcome of the bargaining process. It will not be worth their while to be distracted by the business of bargaining when the matters are of no pressing direct concern and the transaction costs exceed their individual expected advantage. These bargains will not be made. Class action suits can help in these cases, but they are in large part not to be relied upon. Third, there are no parties at the bargaining table for the many species whose existence is threatened by our industrial activity, nor are there parties to bargain for such abstractions as 'pristine nature'. Other parties, such as future generations, don't yet exist.

Elinor Ostrom has studied how communities succeed or fail at managing finite open common pool resources such as grazing land, forests, irrigation waters and fisheries. Research shows that local groups closely linked to the resources in question are often capable of sustainable management and efficient extraction of the products of the commons. In many cases, management is more efficiently organised locally than if rules and infrastructure were to be imposed by external authorities to manage the commons.[16] Ostrom's work shows that the economic model of humans as norm-free myopically short term operators with perfect knowledge of market conditions, focused exclusively on maximizing their individual net worth in monetary terms, is not an adequate model of the complex adaptive systems that comprise real world communities. People are limited in their knowledge, are not wholly rational in their decision-making, are constrained in their choices by cultural factors, are aware of and strongly influenced by social factors such as reputation, and are rarely free of ethical and moral views that may dominate their decision-making. Furthermore, the rule-making that results in sustainable management of the commons is more like an ongoing dialogue between all entities (individual, corporate and government) evoking an experimental chaotic process that, under the right conditions, will move towards sustainable and efficient management of the commons.

Ostrom's experimental laboratory work showed that players of investment games that incorporated the logic of the prisoner's dilemma, when given the opportunity to discuss strategy in face to face meetings between each iteration, tended to regulate their behavior and win group results that approached the optimum. This held even when the behavior of individual players was not revealed to the other players.[17] This robust finding shows that public shaming did not play a role in regulating behavior, but rather some internalized sense of group morality constrained individual behavior, leading to positive group results. In other experiments in which players did not enjoy anonymity and were able to devise agreed upon schedules of sanctions against players who broke ranks to obtain unfair shares, up to 95 % of the optimum group yield was achieved by co-operative play.

---

[16]Ostrom, E. (1999) *Coping with Tragedies of the Commons.* Annual Reviews of Political Science. 2: 493–535.

[17]Ostrom, E., Gardner, R. and Walker, J.M. (1994) *Rules, Games and Common-Pool Resources.* Ann Arbor: University of Michigan Press.

Ostrom concluded that players used complex heuristics, not game theoretical calculations, to determine their actions. When players are able to meet and agree on strategy, and when given the opportunity to devise their own rules for sanctioning rule breakers in games in which players are not protected by anonymity, they spontaneously regulated their individual behavior so that collective results approached optimum outcomes.[18] Players tended to react with great indignation against rule breakers, to such a degree that some individuals were willing to impose sanctions on rule breakers at considerable personal cost, occasionally surpassing the loss imposed by the rule breaking, demonstrating that inherent non-rational psychological factors play a role in these behaviors.

Many instances of local community-based initiatives to take control of open common resources and manage them sustainably have been studied and documented. Reviewing these, Ostrom delineated a set of eight conditions that predict success.[19] In order to ensure sustainable management of the commons, stakeholders should ensure the following:

1. Define clear group boundaries to membership. Increasing the proportion of participants who are well known in a community, and who have a long term stake and reputations of trustworthiness to protect in that community enhances the likelihood that optimal reciprocal behavior will be observed.
2. Match rules governing use of common resources to local needs and conditions. The rules may specify harvesting caps, seasonal restrictions, limitations on the technology used, time of access, and so on. The rules must be seen to be fair.
3. Ensure that those affected by the rules can participate in making and modifying the rules. Research shows that locals are better at specifying rules that actually work. Local groups should be empowered to experiment with rules, which is important for maintaining the effectiveness of the adaptive and continually evolving system. If complex ecological calculations to find carrying capacity are required, for example, it is better to educate the locals than to impose quotas with an authoritarian hand. The latter may lead to rule-breaking and local forces implicitly approving of the rule-breaking.
4. The rule-making rights of community members must be respected by outside authorities. Interference by well-meaning but more distant authorities can break the system. Devolution of authority and explicit support of local decision-making enhances the system.
5. Develop a system carried out by community members for monitoring members' behavior. Informal monitoring among peers will occur spontaneously, but giving it a formal structure will improve the efficiency and adherence to the rules. Permit community members to tinker with the monitoring system to enhance its effectiveness and efficiency.

---

[18]Ostrom, E. (1998) *A Behavioral Approach to the Rational Choice Theory of Collective Action.* American Political Science Reviews. 92(1): 1–22.

[19]http://www.onthecommons.org/magazine/elinor-ostroms-8-principles-managing-commmons 24 May, 2015.

6. Use graduated sanctions for rule violators. Some participants will test the system by breaking the rules and will adjust their behaviour on the basis of the response. Severe sanctions for first offenders can lead to ill-will and can break the system.
7. Provide accessible, low-cost means for dispute resolution. Ensure that higher authorities and avenues of appeal that exist will respect the local decision-making process. Bullying, corruption and political biases can occur at any level. Dispute resolution should be based on clear explicit delineation of rights, by which Coasian bargaining becomes part of the backbone of the system.
8. Build responsibility for governing the common resources in nested tiers from the lowest level up through the entire interconnected system. Keep in mind that the system will be a complex adaptive system that must be able to evolve as conditions change. Evolution of the system should come from the ground up. As far as possible, power should devolve to the lowest possible level, with higher authorities taking educational or mentoring roles in preference to judicial or legislative roles.

We have already noted that human behaviour is not always rational in the way presumed in classical economics. The term 'spite' is used in sociobiology to refer to behaviour that results in greater damage to the spiteful individual than the damage arising from the rule-breaking behaviour of the offending party. How can we explain this behaviour? Similarly, self-sacrificial heroism requires explanation. What does it gain a man to lose his life in battle for the benefit of his fellow warriors? Both spite and heroism can be explained by showing that the loss to the spiteful or heroic individual is more than compensated for by gains among within-group members. Evolution of instincts that drive spiteful and heroic behavior may arise from mechanisms of group selection or of kin selection; not all sociobiologists are in agreement regarding the mechanisms, but all agree that the sum of the gains to within-group individuals in the long term must be greater than the individual losses. This collective non-zero sum drives the evolution of instincts for both spiteful and heroic behaviour. These behaviours are therefore not as irrational as they may appear at first glance. They demonstrate that human behaviour is not solely self-interested in the way classical economists posited.

There appear to be other reasons why behavior is not norm-free myopically short term resource-maximizing strategic play. Most of the highest quality software is open source and free to use. The Internet, for example, runs on a backbone of UNIX that is not proprietary. The open source world is populated by highly talented programmers operating in an informal association, who contribute their time and expertise to the commons with no expectation of direct financial reward. Many of these programmers could earn six digit incomes, which they may forego in order to work on their open source projects. Why? Once again, we must turn to psychological and socio-biological theories to explain this.

The community of open source programmers is held together by their expertise and shared understandings.[20] You cannot fake your skill in this community; skilled

_____

[20]Raymond, E.S. (2001) *The Cathedral and the Bazaar: Musings on Linux and Open Source by an Accidental Revolutionary.* O'Reilly Media.

programmers can read code as most of us read novels. Excellence is widely understood and appreciated, and uncommon cleverness is greatly admired. Talented
members of the open source community can win the approbation of their peers for
their contributions. This is what they work for. It has been argued that this is the
ultimate reward for which we all strive. Why make money, we may ask, except that
we may use it to purchase expensive cars and luxury homes that serve to advertise our
success. Such symbols of material success win us the approbation of our peers in the
entrepreneurial and corporate world. Money is a means to these ends. A talented
programmer can skip the money and go directly to the recognition and status that his
skills earn. Money merely gets in the way. The result is a spontaneous self-organizing
system that arises from the bottom up and manages the many projects in the collective open source enterprise. The open source world is free of the administrative
command structures that characterise the corporate world of commercial software,
resulting in a more efficient self-regulating, organically-growing community comprising many of the world's most talented coders.[21]

Sociobiologists argue that the ultimate reward is access to fecund partners and
greater reproductive success. It matters not whether this is achieved by acquiring
money, fast cars and great estates, or by winning honor and influence in
non-monetary ways. If this is so, then the fundamental basis of economic theory is
mistaken. Perhaps this explains why a great part, perhaps the greater part, of human
endeavor takes place outside the realm of the money economy; the investment that
couples make in each other and their offspring is the most obvious example. But is
this all there is to being human? Are we just chasing reproductive success, driven
by selfish genes that care not what our individual losses may be?[22]

Computer models can be constructed to demonstrate the phenomenon of emergence.[23] By this we refer to properties of systems that do not appear to be properties of
the parts. A widely cited example is found in computer models of ant behavior. These
models can be written with a small number of rules for each ant. These rules do not
enable any solitary ant to behave in a complex purposeful way. In some models, the
solitary ant engages in a random walk until it dies of starvation. However, when many
ants are placed together in the virtual space created by the program, seemingly
intelligent and purposeful behavior emerges from the interactions between them.
Behavioral scientists studying behavior in insects and other species concur that
emergent phenomena are necessary to explain complex chaotic self-organizing collective behavior that is observed in nature. Can we extend the concept of emergence to
human social behavior? Certainly, although it would be rash to claim that we know
what the emergent behaviors are or how to distinguish them from other aspects of
human behavior. Nevertheless, it can be argued that the rules encoded into our nervous

[21]Steele, R.D. (2012) *The Open-Source Everything Manifesto: Transparency, Truth, and Trust.*
Manifesto Series, Evolver Editions.

[22]Dawkins, R. (1976) *The Selfish Gene.* Oxford University Press.

[23]Johnson, S. (2002) *Emergence: The Connected Lives of Ants, Brains, Cities, and Software.*
Scribner.

systems bring about new dimensions of emergent behaviors that transcend those predicted by economic and evolutionary laws of nature. Some people argue that the moral dimensions of human behavior are emergent. The spiritual, we are told, is the component that makes the whole something greater than the sum of the parts.

What about the ethical foundations of corporate capitalism?. We all know that the most magnificent and inspirational mission statements hide many sins. The charter of Enron was exemplary, yet this company gave us one of the most egregious examples of rapacious corporate predation and criminality of the last century.[24] More generally, the advertising industry has fostered a culture of image-creation tantamount to systematic lying, and rendered it ordinary, acceptable, desirable even. Public awareness, itself another aspect of the commons, is corrupted and exploited in the interests of corporate greed. Can international corporations ever possess the emotions of guilt or of pride that may drive moral behavior in individuals? Some argue that a culture of corporate responsibility can arise if senior management sets the standards from the outset.[25] However, many believe not, and in noting this, we can make a distinction between 'free enterprise' that guarantees rights of individuals and 'corporate capitalism' that has evolved into a global device for exploiting these rights.[26] The system has become the birthplace of monsters enriching themselves while destroying the planet and undermining a just and equitable society.[27] It should be cause for alarm to note that none of the mechanisms described by Ostrom operate at the corporate scale. Perhaps it is helpful to quote once again from Adam Smith, the father of economics whose prescient insights were filled with hope and optimism, but also with foreboding and warnings that we seem to have ignored for too long[28]:

> People of the same trade seldom meet together, even for merriment and diversion, but the conversation ends in a conspiracy against the public, or in some contrivance to raise prices.

[24]McLean, B., and Elkind, P (2013) *The Smartest Guys in the Room: The Amazing Rise and Scandalous Fall of Enron*. Portfolio; Reprint Edition.

[25]Collins, J. (2001) *Good to Great: Why Some Companies Make the Leap...And Others Don't*. Harper Business.

[26]Hamilton S., and Micklethwait, A. (2006) *Greed and Corporate Failure: The Lessons from Recent Disasters*. Palgrave Macmillan.

[27]Senator Bernie Sanders (March, 1, 2001) *The Speech: A Historic Filibuster on Corporate Greed and the Decline of Our Middle Class*. Nation Books.

[28]Smith, A. (1776) *An Inquiry into the Nature and Causes of the Wealth of Nations*. Strahan and Cadell, London.

Chapter 2
Early Days: From Personal Awareness to Group Commitment

Jean-Paul Close, Marco van Lochem, Edwin Weijtmans,
Mary Ann Schreurs, Alfred Stein, René Otjes and Hans Verhoeven

2.1 A Personal Story of Awareness and Perception

Chapter 1 develops the historical context of a human world based on economics, trade and consumption mechanisms that evolved into a free and democratic market with political and economic dependencies, rules and relational structures.It shows that we may have reached a dead end as a result of the flaws in this system as they appeared over time up to a level of exponential tension around the world, within the system itself and with our natural surroundings. It concludes by trying to understand the rational and irrational aspects of human group behavior. In essence, this is what it is all about: human beings, their complexity, and the way we manage to progress through time successfully as natural, self-aware, evolutionary, group-oriented creatures. My chapter is therefore dedicated to this wonderful and extraordinary, confusingly evolutionary phenomena, the human species, complex

J.-P. Close (✉)
STIR Foundation, Eindhoven, The Netherlands

M. van Lochem
Odeon, Best, The Netherlands

M. van Lochem
Axians, Eindhoven, The Netherlands

E. Weijtmans
Den Bosch, The Netherlands

A. Stein
Faculty of Geo-Information Science and Earth Observation (ITC), University of Twente, Enschede, The Netherlands

R. Otjes
ECN, Petten, The Netherlands

M.A. Schreurs · H. Verhoeven
Environment, Eindhoven, The Netherlands

© The Author(s) 2016
J.-P. Close (ed.), *AiREAS: Sustainocracy for a Healthy City*,
SpringerBriefs on Case Studies of Sustainable Development,
DOI 10.1007/978-3-319-26940-5_2

beings which I have put at the center of our natural attention. In this story, I use one specimen as an experimental guinea pig, myself (Jean-Paul Close), to try to understand what has hardly been understood before, maybe because of our blind focus on external mechanisms of power, control and submission rather than harmony, symbiosis, awareness and life itself. Can this change?

At the time of my return to the Netherlands in 2001, after an expatriated absence of 27 years, none of the analysis described in this book had been part of my own reality. I had never heard of Ostrom and the commons, nor Kazimierz Dabrowski and his layers of the consciousness, or Matthew Lieberman and his brain research applied to human behavior. This whole world of understanding of our complex behaviour and interpretation of things did not exist for me. Why? Simply because I was not aware, just like the great majority of my fellow human beings is not aware. I had grown up in the world of business development, marketing and hierarchical methods of human resource management, profit and loss, personal growth, etc. I had learned to look at my natural surroundings as if it were an oil painting, enjoying the beauty of nature, experiencing moments of emotional pleasure when hearing the waves of the sea, smelling flowers or feeling the warmth of the sun when on holiday. My sensations were that of an eyewitness, an observer, not of a self-aware part of it all, let alone one with the possibility of sensing responsibilities or the need to harmonize with it. My real world was that of business transactions and competitive careers in executive hierarchies. All this was about to change.

Since my birth in the Den Bosch, The Netherlands, in 1958, I had been living in countries that were in a phase of developing their economies and democratic structures. And I was professionally very active in performing and developing a career within these mechanisms. It never even crossed my mind that there could be risks or consequences to this way of working and thinking. My perception of the world first evolved in the post-war Dutch society of the 1960s, in which the public discourse was about social securities, pension schemes, amount of working hours and taking measures that would prevent a society from entering again into a pre-disposition for public uprising and war. Germany was used as a historical reference while the Dutch society evolved into a "society of caretaking", in which government took over responsibilities from the population with the guarantee of providing wellbeing in exchange for peace.

When I moved abroad at age 16, in 1974, we landed in Barcelona, Spain, where my father had become an expatriate executive for a Dutch multinational. My emotional and rational development evolved further, having to adjust to an English school system populated by the "special breed" of children of ambassadors, executives and scientists, all with a touch of arrogance and financial wealth. Then the Spanish dictator Franco died. This released a tremendous amount of energy and tension for the cause of restoring the lobby for Catalan autonomy, while King Juan Carlos I was assigned to re-establish the local monarchy and national democratic cohesion, originally established by the Catholic Kings in 1492. The subsequent connection with the European Community and the organization of several big events, such as the world soccer championships in Barcelona in 1982 and the Olympic Games and the Expo in Sevilla, both in 1992, caused Spain to enter into its own boost of economic development and real estate bubble. In this ever-growing context, and as far as I was concerned, economic growth was a given, I had not experienced anything else yet. My own career also seemed to evolve in an auto-mated process of professional and economic growth, as if this was the most normal pattern, and one that would last forever. However, at this point I started to question this for myself: "Is this all?"

My first encounter with a national crisis came when I was a student of Mathematics and Computer Science at York University in the UK, between 1977 and 1980, at the time of Margaret Thatcher. This crisis was not yet one that entered my consciousness as something that affected me personally, other than the sudden multiplication of the annual tuition fees by three. Margaret Thatcher has been quoted as saying: "If foreigners choose to benefit from the excellence of our edu-cation then they have to pay for it." My own subsequent professional career had me working in multinationals with world wide executive responsibilities, ending up, after a period in Stockholm, Sweden, back in Spain again, thanks to my knowledge of the local language. My professional career evolved strongly in the field of European computer and telecommunication networks and technologies.

2.1.1 My First Awareness Breakthrough

My first divorce, in 1996, involved my daughter, aged just 18 months. It made me aware for the first time of the inner conflict between moral responsibilities towards my child and the practical complexity of a global executive career. I decided to let go of the latter and develop my professional life around responsibility for the wellbeing of my daughter and harmony with her. The emotional process of vol-untarily letting go of all the benefits and status of being a global executive, without any prospect of financial security after the choice had been made, had a deep impact on me. I had to come to terms with my inner turmoil and the chaos that evolved from letting go of the sense of safety and motivation for external material wealth, learning to trust myself as my own security and resource, no matter what.

Something truly astonishing happened to me. At the deepest point of self-pity and sorrow, there was an inner breakthrough. I had stopped in the countryside near Madrid to take some time to come to terms with myself. For the first time, I observed my natural surroundings with a new sense of harmonic connection, seeing it as a totality of which I was a living and self-aware part. I could sense the colors of the trees and the sky, the flying of the birds and warmth of the sun. None of this had ever broken through into my conscious awareness before, while travelling from hotel to hotel, city to city, airport to airport, as if this reality had been hidden from my senses due to some programmed way of living life. Having gone through the extremely painful experience of letting go and opening up to the realest of realities, a totally new world had suddenly opened up for me. It felt like a revelation. I was not a witness or observer any more; I had become part of it all. This was 1996, and it would mark my life ineradicably from that moment on.

2.1.2   Layers of Awareness

Many years later, I would learn about the theory of the layers of awareness, the levels of positive disintegration, described by psychologist Kazimierz Dabrowski,[1] and the effects of mental growth through these experiences. Dabrowski and subsequent analysts of human behavior describe 5 layers of awareness that all have to do with the "letting go" (disintegration) and the revelation of new levels of deeper insight (integration). Later, I would start adding my own experiences and analysis to the insight, but at that moment, it had become a delight to learn about the working of the conscience, an idea from which I had previously mentally blocked myself when I concentrated only on personal growth in a competitive environment. Stated plainly, "empathy" is a psychological state that requires people to go through a sufficient process of disintegration so as to become capable of understanding and valuing the pain or beauty of another. Curiously, analysts state that only a small minority of people reach that state of awareness, while the vast majority remains stuck in the lowest two levels, just worrying about competitive growth (I want a bigger car than my neighbor's) and survival (as long as I can pay my mortgage and go on holiday twice a year) without even sensing other issues at hand. This could explain why the human world has such difficulty responding to climate change, pollution and the many other global issues. We observe it as witnesses, as outsiders, so that it does not reach our inner selves, so that it does not affect our daily routines and choices around short term self-interests (Table 2.1).

> Interestingly, we can consider that our personal and individual human development may reach level 5 of understanding of the genuine principles and complexities of life. We do this through structural processes of letting go, with awareness breakthroughs emerging out of our own mental chaos. But our organized structures, such as government and business, have historically evolved only to levels 1 and 2, disintegrating completely in conflict and chaos while showing an extremely slow capability for breaking through into collective awareness. In my perception, we have been struggling collectively to let go of levels 1 through 3 for over 5000 years now, having structured society around egocentric greed, hierarchy and control ever since structures were needed. Levels 4 and 5 may now lay ahead of us, possibly producing an historical breakthrough and evolutionary leap for our species. This is where I position this documented exercise, as an invitation for people and institutions to join together into a totally new and deeper societal reality.

Soon, my own world, within the energetic positioning of the administrative and political center of Spain, Madrid, evolved again in the development of my own financial security. The next few years were rollercoasters for me, veering between the external and inner securities of extreme entrepreneurial success and financial growth, sudden collapse and rebuilding the cycle all over again. As a consequence, my emotional views of economic reality and the working of the markets became intensified with empiric experiences that I, at the time, could not yet rationalize or describe as was done in part one, above. The collapse of the internet bubble in

---

[1]Dabrowski, Kazimierz, Andrzej Kawczak, and Michael M. Piechowski. *Mental growth through positive disintegration*. London: Gryf Publications, 1970.

Table 2.1 Dabrowski levels of disintegration and integration compared to J.P. Close cyclic complexity

No	Dabrowski level	Dabrowski characteristic	Jean-Paul Close characteristic
5	Secondary integration	Harmony	Harmonic wellbeing and co-creative action
4	Organised multilevel disintegration	Self organisation	Awareness breakthrough and personal action
3	Spontaneous multilevel disintegration	Ideal versus real	Awareness development (doubts between what is right and wrong)
2	Unilevel disintegration	Conflict	Competition and chaos
1	Primary integration	Self-gratification	Growth
0	Not defined	Not defined	Conception (=5 from previous cycle)

2000^2 meant the definitive end of an era for me as a telecom professional. In just one month, two of my major contractors and sources of income went under. It became clear to me that I would have to open up to new realities, including professionally. I also became more clearly aware that life does not consist solely of phases of growth and conflict. There is much more to it and I was determined to figure it all out. But first, I had to modify and adjust my own life's patterns.

I decided to move back to Holland with two motives:

- I perceived the Netherlands as the geographical center of executive Europe from a regional multinational point of view, creating better perspectives for a future for someone with my executive profile;
- I wanted my children, the second one of whom was on her way, to have the benefit of the sort of multicultural foundation I had enjoyed and valued as important, including the Dutch culture.

2.1.3   My Second Burst of Awareness

Upon arriving back in the Netherlands, after the already-mentioned absence of 27 years, I was immediately struck by the intense culture changes that had taken place in my country. At the time of my departure in 1974, there had been a sense of national unity of purpose, a shared vision of a new society striving for social justice and peaceful fulfillment of the material and immaterial needs of every citizen. What I found, however, was a harsh money-driven, consumption-orientated and

---

[2]Fransman, Martin. *Telecoms in the Internet Age: From Boom to Bust to?* Oxford University Press, 2002.

speculative culture, with a tremendously bureaucratic and prejudiced, even dis-
criminatory, government wielding its clumsy and insensitive regulations. This was
not the country that I had been returning to in my mind.

Never, anywhere in the world, had I encountered problems finding housing for
me or my family. In the Netherlands, there was a waiting list of 5 years for a social
rental home. Options were available in the free rental market, but this was at
extreme prices adjusted to the expat market of temporary housing, not the rootless
people "coming back" after an economic crash in the world of expats. Another
option was to purchase a house. This was conditioned on the collateral of indefinite
labor contracts and extreme prices that had been manipulated in the real estate
boom since the '70s. A house that was purchased in 1970 for the equivalent of
€15,000 would sell in the year 2000 for €250,000. Paying for the house four times
(the original price and the equivalent of 30 years interest rates at 7 %, representing a
doubling rate of 10 years!) still left a perceived profit (not real, because the profit is
relative to the past not the present) of €190,000. What justified this huge value
increase? What justified the extreme profits of the banks? Banks would draw the
same line into the future to justify their 120 % mortgage offer against 5 % interest.
The same house would be worth, according to them and the common belief, 3
million € in 2030. It became clear that such economic bubbles of perceived
enrichment, without doing anything in terms of productivity, and an unnatural
30 year mortgage claim against future labour, had become a common problem of
collective blindness, including mine. But where can I live with my family if the
environment is as harshly manipulated as it is? We had no choice. We bought a
house.

My second boost of awareness came when my family, consisting of 5 individuals, 2 adults and 3 children, had to apply for residence permits on an individual
basis, regulated by the system's bureaucracy. We were not treated as a family, while
in "my world," a family would be considered the basis of society. Government
reasoning was related to the level of abuse by people entering the country and
enjoying the socio-economic benefits through fake marriage arrangements. My
second wife was Brasilian. Each family member was emotionally tormented by the
possibility that one or more members of our family would not be allowed to remain
in residence. This caused so much fear and insecurity that the family cohesion
started to suffer. My wife was forced to "integrate" into the Dutch culture, valued
against materialism, language and having access to the labor market, rather than
developing her perception of ethical value: family cohesion. The Dutch demands
were contrary to her and my cultural beliefs of family harmony. The system's push
was sensed as inhumane, immoral and unjust. She felt so much stress and
aggression against her own inner values that she reacted back with aggression
within the family, with natural evasive behavior of escapism.

After one year, the stress had become so great that the family union broke
up. I had to take instant responsibility for my children by going into hiding against
the aggression. My wife fled the country, leaving me to carry on as a single father
for the children within a totally disrupted community. We lost our house, our
income and stability. Thanks to the help of family, I was able to survive, but the

general overall governance culture necessitated that I again become a participant in the tax-paying community, subcontracting, if desired, the stability and education of my offspring to the system for the sake of money. We were made marionettes in a money-driven reality. If I refused, I was not provided access to social security; if I agreed, I needed to accept the vulnerability of my children in the face of possible abduction.

I refused.

This is not the type of society that I want to pass on to my children.- J.P. Close (2005)

My first burst of awareness had made me aware of my deep inner sense of harmony with my surroundings and my responsibility for living life, trusting my abilities instead of blindly following external securities and rules. My second burst of awareness reconfirmed this and made me conscious of society as a simple (no matter how complex) set of rules measured against a diversity of possible values that can go from human securities and cohesion (post-war development) to trying desperately to sustain an artificial system (money) at the expense of what it was built for in the first place. My second boost of awareness was related to the way communities develop and change "polarity" from cohesion to greed, from unity to falling apart. My awareness, as experienced at the level of family cohesion, could also be applied to the larger community. It became clear to me the way in which group patterns appear, growing up to a limit and then tending to collapse when they reach a certain point (point of singularity—see Fig. 2.1). It dawned on me that this happened in living nature all the time. In my personal view, Dutch society was on

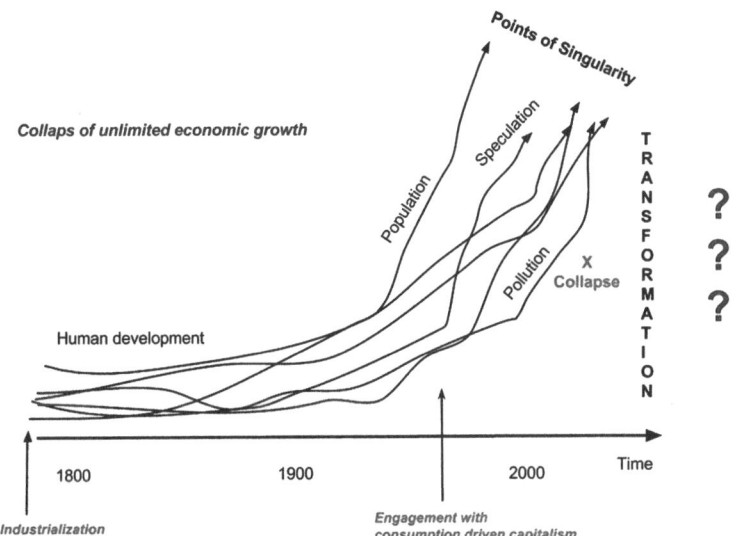

Fig. 2.1 The engagement with consumption driven capitalism produced the misconception of seemingly unlimited economic growth

the verge of collapse, sustaining itself only through artificial measures of inhumane proportions. It had grown from a social democracy into a self-imposed financial dictatorship. I did not feel part of this and was surprised that the population did not react. The situation could not last, because it lacked all sense of community. Why was there so much blindness? Why so much apathy? People only complained. Why so much dependence? Why so much fear?

One of the interesting consequences of a second positive integration described by Dabrowski is the gradual disappearance of "fear" as deeper awareness connects with the essentials of life and the harmonic connectivity with social and natural resources. No fear is needed. One learns to possess nothing so there is no fear of losing anything. One becomes humble, free and fearless. Ironically, for those competing in life this humbleness is seen as a weakness, while in reality, it is the most powerful and lasting of all layers. All points of singularity feared by so many then become obvious turning points, starting points of a new beginning, not a fearful dot to be avoided all times, but one to be cherished as essential for life renewal and evolution. While the entire world was either blind to this or trying to avoid collapse, I found myself engaged by motivation, joy and determination to define "What's next?"

2.1.4   Defining a New Society for Myself

My own sinusoidal life pattern resembled the cyclic wave analogy described by economist Kondratiev.[3,4] It could even be related to the work of Pythagoras and Galileo Galilei on musical patterns.[5] Musical strings contain a secret that explains patterns of life that can be traced into the formation of huge constellations all the way down to the positioning of planets in relation to their solar system, the way our weather behaves and the alliance of molecules to form life and evolve through DNA strings. For the first time in my own life, I could observe the uniting powers of harmonic rhythms in nature, the fractal growth patterns in life, including the phases of collapse and the analogy in economics, group dynamics and business develop-ment. All this knowledge and awareness was a personal privilege that became difficult to share with others. What had become very real for me was abracadabra for nearly everyone else. When I started talking about it, my audience's attention would fade quickly, rapidly reverting to their daily issues. They would admire and acknowledge my fearless single fatherhood and the perceived challenges this

---

[3]Goldstein, Joshua S. "Kondratieff waves as war cycles." *International Studies Quarterly* (1985): 411–444.

[4]Rainer, M. E. T. Z. "Empirical Evidence and Causation of Kondratieff Cycles." *Kondratieff Waves, Warfare and World Security* 5 (2006): 91.

[5]Walker, D. Perkin. "Kepler's celestial music." *Journal of the Warburg and Courtauld Institutes* (1967): 228–250.

brought me, but not the reasoning behind it. Nor did people understand that I was busy and highly motivated all day, and every day, but did not "work" for a boss.

The mainstream human's reality was based on external material securities, managed through a competitive system of labor dependence and status that still had a strong relationship with the old industrial era. Work and living life were seen as two different things, a dual inside and outside system which perhaps clarified why people might see their family life as something separated from society. The fact that I considered family life and my contribution to society as being the same thing meant that I perceived "human values" in a different way than those surrounding me. They calibrated family life against the level of income generated by this other life, called "work". One either had work or not, receiving an income either out of labor or social security. The binary switch between work/no work was just income-, activity- (to Do) and status-related, it did not challenge people to open their minds to broader realities (To Be). The "To Do" and "To Be" entered my curiosity (Fig. 2.2).

None of the above experiences would have entered into my awareness either if I had not returned to the Netherlands to witness the dramatic change that had taken place between 1974 and 2001. I would have mainstreamed my life like anyone else. Now, it had reached my understanding in all its complexity and it had a huge impact on me. Even my return to Holland at age 43 after an absence of 27 years could be placed along the cycle of the musical resonance and the vibrating string theory of nearly 54 years, equivalent to a full Kondratiev cycle and the famous 7 + year sub-cycles of ups and downs. Was this casual? Or part of my own life's pattern within a much larger symphony of patterns that we are normally not aware of? What significance does "harmony" have as opposed to growth? What role does money have in all this and human awareness? Is harmony the status quo or hard work? If I can realize how it works for one single human being, with all their moments of stress, pain, new awareness and new phases of inner and external harmony, how will it work when we consider masses of individuals in a lump sum?

A new world of investigation had opened up to me.

Fig. 2.2 The way we tend to perceive ourselves, influenced by society's rules and culture

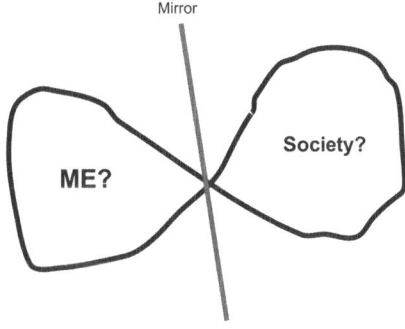

2.1.5   Key Human Values

In my own two different occasions of boosting of awareness, I had to make my own mind up about ethics and responsibilities without relying on rules and regulations. My mind had opened up to see that the ethics of the societal system that demands solidarity with economics and social security is totally different from the ethics and sense of responsibility that nature imposes upon us. On both occasions, I reasoned that neither status nor employment is relevant to a good life if it comes at the expense of our future generations. What would you save first in times of war? Yourself and your children, or your possessions?

> In our dual lives (work and life) we had separated responsibilities, developing our own at home and letting governance take care of the rest. The common denominator had become "money dependence" and not our key human values.

It became clear to me that, as individuals and as a society, we had inverted our priorities. Key values of human evolution can neither be delegated nor purchased. They represent responsibilities that we carry alone and together. Responsibility cannot be expressed in money, nor can life, which is too valuable. The true human values resonate and create cohesion and commitment; if neglected or destroyed, they cause communities and life itself to fall apart. This is how I had experienced it in my own life's evolution.

From my new point of view and awareness, the analogy of musical bonding, that keeps constellations or molecules together in living patterns, applies to societies too, starting at the family level and expanding into the entire community, including the meaning and operational reality of institutions. Recently, I looked back at my inner discoveries, motivated by the writing and editing of this book, and came across the work of brain researcher, Dr. Matthew Lieberman.[6] He explains in an impressive TED talk that "Abraham Maslow's hierarchy of needs" was wrong. The first basic essential of every human being is social cohesion, not the fulfillment of our primary material needs (Fig. 2.3).[7]

This simple remark, which I called "the Lieberman correction," has huge consequences if taken to heart in establishing communities and societies. Today, we organize ourselves around the idea of ensuring the abundance of basic needs in a consumption-driven society. When people have what they need, then there is no more need for social interaction. And even worse, when people fear losing what they need in such a caretaking consumer environment, the psychological tendency is to avoid social interaction even more or become aggressive towards one's surroundings in a primary reaction of defending self-interests (hoarding). We take the liveliness and creative purpose away from the individual and the community,

---

[6]Ochsner, Kevin N., and Matthew D. Lieberman. "The emergence of social cognitive neuroscience." *American Psychologist* 56.9 (2001): 717.

[7]Rock, David. "Managing with the brain in mind." *Strategy + business* 56 (2009): 1–11.

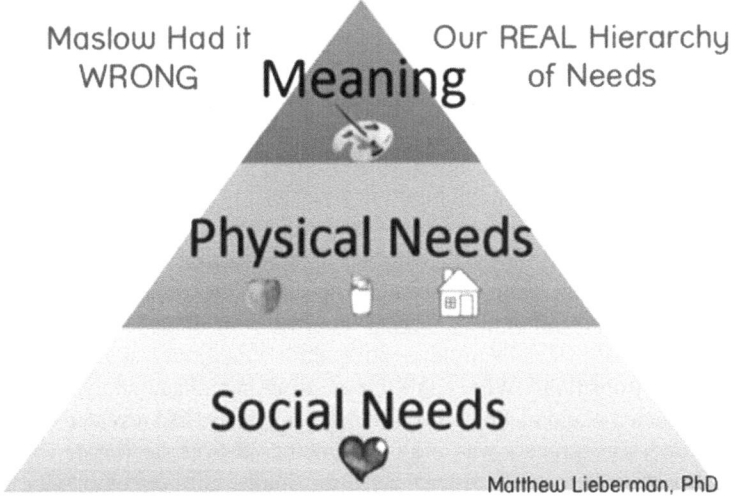

Fig. 2.3 Dr. Matthew Lieberman, Maslow was wrong

resulting in an unproductive "frozen ego"[8] type of situation. If we wanted to revert back to key human and evolutionary values, I would have to rephrase society around such importance of cohesion:

- Society needs to collectively respect and facilitate human values, such as health and safety, above political and economic interests, just like we do on the family level;
- Society needs to engage in harmonization of our human and natural surroundings by taking proactive responsibility instead of delegating it into a remedial system.

My personal decisions in 1996 and 2003 had been intuitively done with this in mind. Now, I learned to interpret and rationalize it, with important references all the way back to the times of Pythagoras. Apparently, we had needed 2500 years of societal trial and error at the collective levels 1 and 2 of Dabrowski's 5 layer awareness scale to reach a collective point of singularity of blind unethical growth, to subsequently crash and open up to the levels of collective positive disintegration rather than just the individual.

2.1.6   Inside = Outside

I had felt that the key responsibilities of any self-aware adult human-being towards their own selves and the children in their home should remain the same when

---

[8]Close, JP (2013) "Frozen ego's" when apathy takes motivation away to do something other than complaining.

stepping outside the front door. Why should the priorities in society differ from home? Did my own consciousness differ from that which surrounds me? And what did I need to do to establish the synchronicity? Accept the situation that got me as part of a family union into trouble and adjust my commitment to life to that which was imposed on me by a "wrong society"? Or should I apply the lessons of life learned and try to introduce a new practical reality based on the productive energetic patterns of harmonic or symbiotic relationships? I decided that the latter gave me a better sense of purpose, as I could take responsibility both for my choices at home and in the development of my professional or social activities outside.

My inner quest became to understand why groups of people connect and come alive in productive communities, such as families, a business community or entire societies, and what makes them fall apart again in crises, recessions, bankruptcies, divorce or confrontations. What could I learn from nature and apply to society? A new experimental world in which I could look for answers had revealed itself. And I was not alone. This process was and is happening all over the world. The Ostrom experiments described in Chap. 1 are a clear example, but our complexity is much larger than a bunch of individuals. Theory is now abundantly available, but putting it into practice is severely handicapped by many influences and a lack of leadership.

2.1.7   To Be and To Do

In 2005, my trust in government was less than zero due to the way it had treated me and my family, and the painful consequences we had suffered. A local government should protect and enhance local values, never become a danger itself to its community and surroundings. In hindsight, I am now grateful, because it had opened my eyes to a diversity of realities, and it had also opened my mind to intense complexities and tensions that could be explained by natural psychological patterns of conservatism and managing security, alternating with forceful or voluntary periods of leadership and intense change. In 2005, the overall national pattern was focused on economic growth with a strong democratic push to conserve social securities and a local perception of wealth. There was no overall sense of need for change in the country. Any reference to the lack of symbiotic resonance, broken harmony and risk of crisis was waved away. So how did my personal breakthrough relate to what I wanted for the country?

When I was invited by my friend Prof. Paul de Blot[9] (Business Spirituality) to give a guest lecture at the Business University Nyenrode, a new puzzle piece fell in place. Paul explains human awareness development along two lines: that of what one does (To Do) and that of what one learns as a consequence, developing what one is as a person (To Be). There is an element of trial and error. We experiment

---

[9]Prof. Paul de Blot–Nyenrode http://www.pauldeblot.nl.

with our actions and interpret the results. That's how we learn to value our senses, distinguish between safety and danger, remember things and proceed with new experiences. Our earliest concerns when born are food and protection. When we grow up, we encounter competition with others while in search of our genuine selves. We enter into conflicts, win and lose, getting to value both, up to a point that we learn to avoid conflict by enhancing what we are, the uniqueness and authenticity that needs no conflict. We then establish harmonic relationships to have children of our own. It is a natural process (Fig. 2.4).

In our discussions, I presented my own view of an inner breaking point, the turning point when To Be starts to lead To Do. While we are growing up through trial and error (we do and learn to be), we may encounter a unique moment of inner revelation after which the trial and error disappears and we develop a creative type of empathy with our surroundings and the need to harmonize. When we do things, we do them with a harmonizing reason. We start to contribute instead of just take.

My own focus on creating harmonic relationships and shaping communities had to find a productive way forward. I had lost confidence in my fellow citizens, whom I found to be blind consumers and workers without any notion whatsoever of consequences, and their own democratically chosen government, dedicated to fulfilling that public desire while raising taxes to remediate the damages. So I developed my transformative mission in the only remaining area where I could try to find enough ground to make a difference: business development. I started coaching businesses and business transitions. In entrepreneurship, I still hoped to find real potential to change society through innovation. Just think of the effects on society of microcomputers, software and the appearance of the Internet. Maybe I too could pull enough entrepreneurial strings to make a breakthrough somehow (Fig. 2.5).

Fig. 2.4 Our natural self-learning process through trial and error

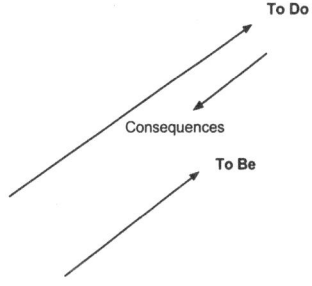

We learn through backwards interpretation

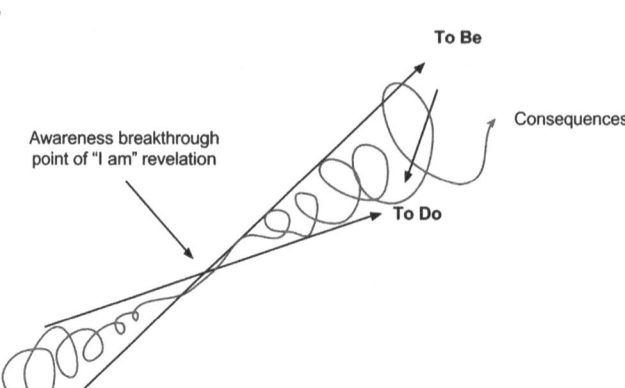

We evolve through forwards interpretation

Fig. 2.5 The awareness breakthrough point defined by Jean-Paul Close

2.1.8   Welfare or Wellbeing?

If I wanted to understand my own inner development and learning processes, I realized I had to write them down and share them with my surroundings. Having arrived at my own insights about the harmonic and symbiotic relationship with my surroundings, how would I apply this to positioning a type of entrepreneurship? Having studied International Business Studies at Nyenrode in The Netherlands, completing two Spanish masters' degrees in marketing and business management, respectively, and after multiple business initiatives of my own, including the one that earned the denomination of Best Business Idea of the Year in the '90s, I had a pretty good idea of the evolution of business over the years, especially since the beginning of industrialization in the 18th century in the UK, as well as pre-industrialized processes and trade in the Netherlands since the 15th and 16th centuries. In fact, a clearly differentiated pattern appeared between the way welfare developed through speculative risk-taking in competitive trade channels with commodities and luxury goods and the planned symbioses with which farmers produced basic human needs in close relation with nature and the seasons. The Dutch developed a golden age around speculation, establishing risky trade with the entire world, especially Africa and the Far East. It is interesting to see the two different worlds of entrepreneurial success expressed in semantics. The Dutch language has two words that, in essence, mean the same thing: "to be successful". They translate as follows, both having totally different backgrounds and underlying reasoning:

- Welfare: as in "welvaren," the Dutch word for wishing sailors a safe and successful trade journey, similar to the hunting groups of the ancient tribes who risked their lives and health to hunt for food. "Welvaren" means literally 'good sailing.' A person who is referred to as "welvarend" is considered rich and successful.
- Wellbeing: as in the Dutch idiom "goed boeren," meaning the finding of a balanced and cyclically productive relationship with the natural surroundings for food production. "Goed boeren" means literally 'good farming.' A person who is referred to as "boert goed" is equally considered rich and successful.

In all societies, we still find both lines of thinking in the development of the countryside and the way people interact with each other. Since the process of industrialization, the entire business industry had developed around fragmented expressions of welfare, measured in financial benefits, not in wellbeing. Trade, speculation and industrial productivity had a natural attraction for people who also desired to consume the luxuries that were produced. It became the impulse for the development of cities in which dynamics of logistics, industrial productivity, profit, infrastructure, growth and the availability of skilled labor were key factors for competitive success.

At the end of the 18th century, pollution in cities was so serious and the mortality among the working force so high that governance was needed to try to establish the wellbeing required to sustain a living community rather than a dying one. The first constitutions were formulated to address such imbalance by introducing rules, regulations, bureaucracy, compulsory education and controls. Welfare and economic growth remained dominant, while wellbeing became a reactive, remedial second priority. As globalization has evolved, we see that the tension between welfare and wellbeing has grown to unsustainable proportions. We now realize that we need to turn those priorities around by putting wellbeing first, with welfare as a sub-system that sustains the development of wellbeing. This turnaround would normally be done through a natural process of collapse. I had proposed a voluntary process of awareness development, breakthrough and self-organized transformation. This became my personal mission. "To Be" (wellbeing) had to start leading to "To Do" (welfare), at the societal level as well. Was society approaching the afore-mentioned breakthrough point?

2.1.9   Business Transformation

In 2005, the early signs of a forthcoming (economic) collapse were clear but found hardly any ground for structural attention, as welfare mechanisms were still booming. Trying to act with some expectation for a positive result, I placed my focus on change by redefining entrepreneurship. I summarized the (r)evolution of business as:

The business transformation from using the planet and people for financial benefits (welfare) into serving the planet and people for sustainable human progress (wellbeing).

This change was characterized as the evolution of business spirituality[10] of the 21st century and resulted in the 5 keys for business success (5K method), published in 2005[11] and 2008.[12] The term "business spirituality" often confuses people who relate "spirituality" to some sort of metaphysical dimensions or religious dogmas. Spirituality in this sense refers solely to the search for inner meaning of an entrepreneurial mission, a purpose that is powerful enough to connect the people participating within a level of creativity and productivity that goes beyond that which one would expect from them. The Business University of Nyenrode had a chair studying this, and these ideas brought me back in touch with Professor Paul de Blot.

The major change envisaged for business development was the challenge of taking entrepreneurial responsibility when addressing the global issues that concern us as a global community. Examples of those issues are carbon dioxide emissions, climate change, pollution, destruction of landscapes, huge migrations, financial manipulation motivated by greed, speculation, destruction of natural resources, etc. (Fig. 2.6).

As I worked as a consultant and coach during those years, I realized this particular entrepreneurial ideology could be tested in the boardrooms of many local and multinational organizations. Various business plans were subsequently written with the use of the 5K[13] and 4 × profit[14] methods as sources of inspiration. The money-driven resonance in the executive boardrooms of the businesses that I coached was, however, taking most of the attention away from the key values that authenticated the original business proposition at its conception. Speculative growth (welfare) and survival (competition) prevailed over symbiotic content (contributing to wellbeing), placing the organizations in short term aggressive battles for survival with hardly any long term vision. A 5K based indexation study of over 300 enterprises in the Netherlands in 2007 revealed two interesting conclusions:

1. Enterprises born after 2000 were much more aware of servant needs, positioning themselves much more strongly in the field of responsibility for wellbeing and co-creation than the older companies that concentrated on speculative welfare and self-centeredness.

2. The overall average of the business indexation of the analysis was a "C," meaning that business in general did not contribute to the welfare development of the Netherlands anymore, despite its focus. Global speculative competition

---

[10]Wilber, K. (2001) *A theory of everything: An integral vision for business, politics, science, and spirituality.* Shambhala Publications.

[11]Close, J-P. (2005) *Handboek voor de (toekomstige) Marktleider.*

[12]Close, J-P. (2005) *Succesgids voor Ondernemers.*

[13]5K = five keys to entrepreneurial success in the 21st century: Key 1: Market definition, Key 2: Positioning, Key 3: Market perception, Key 4: Communication strategy and Key 5: Management capacity.

[14]4 × profit is called the Pyramid Paradigm (Close 2009) and refers to the 4 profits of values-driven entrepreneurship: profit for the customer, profit for society, profit for the environment and profit for the company, "profit" being a synonym for "added value".

Fig. 2.6 Business transition

Business Transformation

Traditional entrepreneurship	Multidimensional entrepreneurship
Product oriented	Value oriented
Materialistic	Purpose driven
Competes on price	Competes on change
Hard	Hart
Money is goal	Money is a means
KPI Managed	Vision managed
Growth	Content
Personnel works	Personnel contributes
Optimization	Innovation
Individualistic	Cooperative
Short term	Long term
Follower	Leader
Distrust	Trust
1 x Profit	4 x Profit

Copyright: Jean-Paul Close

had crossed a line of saturation that made economies rise at the expense of stability. This also meant that business in general was taking values away in a destructive manner. A major crisis was only a matter of time.

While the need to balance social, ecological and value driven economies became a general coaching argument, a global movement appeared under the denominator "People, Planet, Profit." Both philosophies, the 4 × profit Pyramid and the 5Ks compared to the PPP views, presented similar lines for progress, even though the international entrepreneurial description of profit was still predominantly referred to in terms of financial gain. My own definition of profit was expressed in terms of sustainable progress through the benefit of genuine human and ecological value creation. This distinction between money and value became very important. During 2008, a few analytical books on the 5K multidimensional indexation of specific industries (supermarkets, banks and waste management) temporarily popularized the index, but also showed the dramatic state of the Dutch economy, society and lack of value-driven entrepreneurial spirit due to short term money-driven focus on survival. 5K Consultancy became more of an entrepreneurial service of painful criticism than support. The credit crisis in 2008 did not come as a surprise. In fact, it was welcome proof of insight that had been neglected for years by the Dutch and international financial institutions, business development and government. The subsequent massive capital injections into speculative banking, the Arab Spring as spinoff, the many different crises everywhere, the growing worldwide instability, and migrations as a result of natural and human-made catastrophes were a logical proof of the vision and awareness but also the persistence of global "leadership" to sustain the old paradigm at any expense.

2.1.10   Leadership Versus Management

The 2009 book[15] in Dutch "Secrets of True Welfare" introduced the model of human and natural complexity that had been developed by deliberately trying to combine moral human complexity and ethical awareness development (TO BE) with the complexity of organizing human communities (TO DO), as explained earlier from a breakthrough point of view. The two lines were now plotted as two orthogonal lines in which both our organizations' structures evolve as well as our ethical understanding of ourselves.

A crisis is simply seen as the process of letting go of an obsolete past that gave a sense of security but reached a level of unsustainable progress. A crisis always brings in two psychological lines of action:

- Management: that tries to develop and maintain the past in an attempt to grow or optimize what was a cash cow or restore what threatens to be lost.
- Leadership: that accepts the breach and looks at ways to restore harmony by introducing adaptive innovation and change using new levels of awareness.

In The Netherlands, we could strongly sense bureaucratic management dominance at all levels of society. The past, based on welfare development, had been so rich, so wealthy and relatively safe that the entire democratic structure wanted to restore that kind of welfare and return to that past. Capital injections were applied and the bureaucracy enhanced to avoid change (the red line in the Fig. 2.7). Leadership (the green line) introduces changes that upset the structures of the past, which need to be replaced by modern interpretations of reality. But old structures have a lobby, an importance in the old infrastructure, and will pull other organizations along if chaos arises. A culture of avoidance and fear builds up, with tension between management and leadership. This starts with the dominance of management (the red arrow). Gradually, it becomes influenced by emerging leadership, either because the crisis is so strong that management cannot deal with it or because leadership's propositions get so much support that a change of sides must follow.

The credit crisis opened everyone's eyes (first awareness breakthrough) to the prospect that a total collapse could be expected and that this could be avoided temporarily and delayed only through artificial measures. Consultancy did not help anymore. To contribute to society, another vehicle was needed. In 2009, the STIR Foundation was launched. In my view, the cosmetic changes within an economy would not work anymore. We had to change things completely. Business development in the material world is in crisis due to misuse of our environmental and human resources. We needed to replace the central position of banks and money with something of much greater importance: the human being. If we exchange money-based welfare for human-based wellbeing, everything would change (Fig. 2.8).

---

[15]Close, J-P. (2009) *Geheimen van echte welvaart.*

Leadership   vs   Management

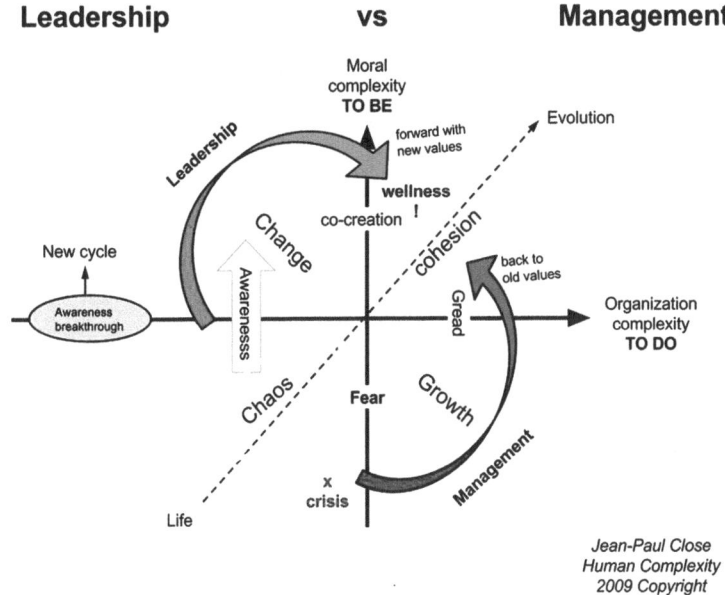

Fig. 2.7 Moral versus organizational complexity (color figure online)

Fig. 2.8 The human being as central given in the City of Tomorrow (This first national logo shows an artistic approach, placing the human being centrally in the text "Stad van Morgen" (City of Tomorrow)—this logo was a contribution to STIR by De Heeren van Vonder in Eindhoven)

2.1.11   STIR Foundation—City of Tomorrow

The foundation had the objective of positioning itself in the field of awareness-based co-creation towards key human values of wellbeing, a drive for transformative change that affects the entire society. Wellbeing, in terms of

Fig. 2.9 The international STIR logo (The international logo of STIR shows the awakening STIR consciousness in the center, surrounded by balancing spiritual, emotional, physical and rational awareness) with consciousness as learning vehicle

harmonizing society around the evolutionary aspect of "sustainable human progress", was defined as (Fig. 2.9):

> **Sustainable human progress** is to keep working together on a healthy, vital, safe, self-aware and self-sufficient human society within the context of our ever changing natural surroundings.[16]

This definition served the STIR mission better than the 1987 Brundtland[17] definition put forward by the United Nations:

> **Sustainable development** is development that meets the needs of the present without compromising the ability of future generations to meet their own needs.

STIR could take personal and institutional responsibility for wellbeing-based harmonization of our present time, as well as our responsibility for the wellbeing of future generations. Having defined our common sustainable evolutionary focus, a new phase in our democracy could be announced. We had no more need to debate our direction, because that had been fixed by the sustainable human progress and wellbeing definition. We could engage by synchronizing our decisions and priorities to this definition rather than spending time democratically disputing the direction from self-interested points of view, often prioritized by our dependence on money rather than our ability to create true measurable values.

2.1.12 Sustainocracy

A new model with which to calibrate our societal structure was born. It distinguished itself in name, commitment, energy and mission from the old democracy. For a long time, liberty, as in the sense of 'democracy,' the participative ability to vote for decision-makers, connected the powerfully desired freedom of speech and

---

[16]Geheimen van echte welvaart—Close (2009).

[17]Redclift, Michael. "Sustainable development (1987–2005): an oxymoron comes of age." *Sustainable development* 13.4 (2005): 212–227.

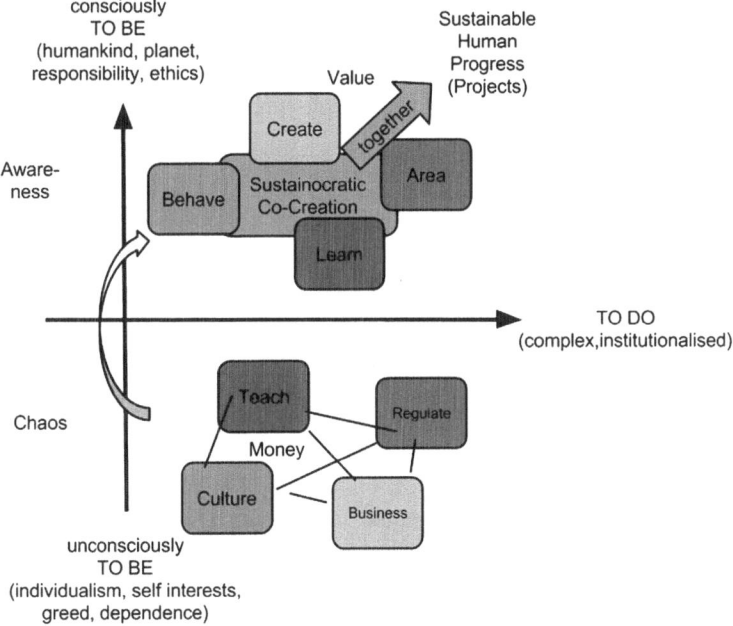

Fig. 2.10 The transition to co-creation based on awareness

choice in times that this was not common practice. Now that we have this sense of liberty, we find that communities tend to value personal securities as a common right and develop hierarchies of greed using democratic processes that do not include productive patterns to sustain those rights. The current democracy became firmly anchored in the engagement with the money-driven welfare structure and individual desires for financial growth as a perceived form of social security.

When it became clear that this democratic structure had destructive consequences, apparent during the current period referred to as the Anthropocene,[18] there was no way to deny honestly that an alternative was required. We needed to reflect on our actions and redefine who we are as a species. With 7 billion living participants in our species we could not afford to learn through our mistakes anymore. Sustainocracy invites us to engage in sustainable human progress by defining and accepting concrete human values, key responsibilities and ethics that have appeared in our mind while suffering the consequences of our old behavioral patterns and governance (see Fig. 2.10).

The term **Sustainocrat** was introduced to represent sustainable *human* progress with which to connect. It is physically represented and symbolized by a real (independent and free) human being, a symbolic function that engages all possible instruments in the sustainable progress definition. This human being symbolizes the

---

[18]The term "Anthropocene" was introduced by Alexei Pavlov in 1922.

value-driven evolutionary essence of the species, replacing the money-driven image of a bank in the old system. With the contrast between two different paradigms now defined, visualized and personalized through positioning of the Sustainocrat, the mission could be unfolded into a new societal complexity.

2.1.13   City of Tomorrow

The STIR Foundation's mission rapidly received a new name from its early participants in 2009. They started referring to the activities as the "City of Tomorrow". STIR found an initial positive entrepreneurial setting in the new International Center of Sustainable Excellence (ICSE) in Eindhoven. This ideological center was positioned to help develop awareness through conferences and a permanent exposition of enterprises that had a story to tell or a product to show in the context of some explanation of sustainability. City of Tomorrow initially seemed to fit well among other inspiring initiatives, such as The Natural Step,[19] Cradle to Cradle,[20] Biomimicry,[21] Earth Charter,[22] etc. Surrounded and interacting with dozens of value-driven initiatives in the field of "sustainability,"[23] STIR organized congresses and a large variety of purpose-driven working groups. Issues like energy transition, CO_2 emissions, healthcare costs and transformation, sustainable housing, city quarter transformation, mobility, the new way of working, leadership, healthy schools, education, pollution, etc., were addressed. However, the obstacle of the money-driven mentality of participants, and a transaction-based economic reality, reversed responsibilities. As soon as an initiative was close enough to start up as a pioneer, the arguments about ownership, business concept and profit allocation would break up the alliance. Self-interest still prevailed for the sake of individual survival in a society that had been hijacked by banks through long term public and private[24] mortgages and debt structures. Most people had important financial obligations to deal with every month and forcefully ran a short term survival scenario without time, room or support to fulfil a more complex leadership vision and mission. Only people that could break free from such burdens could connect to the leadership initiatives of Sustainocracy.

---

[19]Robèrt, Karl-Henrik, and Ray Anderson. *The Natural Step story: Seeding a quiet revolution.* Gabriola Island: New Society Publishers, 2002.

[20]McDonough, William, and Michael Braungart. *Cradle to cradle: Remaking the way we make things.* MacMillan, 2010.

[21]Benyus, Janine M. *Biomimicry.* New York: William Morrow, 1997.

[22]Earth, Our Home. "The earth charter." *Worldviews* 8.1 (2004): 141–149.

[23]No single definition for sustainability existed. Every organization and individual defined it in their own way. The most common interpretations were around energy (e.g., solar panels) and material resources (e.g., Cradle to Cradle). Not many people referred to the vulnerability of the human being, but mostly to the risks to our well-being and luxuries.

[24]Debelle, Guy. "Macroeconomic implications of rising household debt." (2004).

The business case of ICSE suffered, and the organisation went broke within one year of its announcement; however, it left a lasting impression in the City of Tomorrow. The short period of the ICSE was significant for three reasons:

1. It brought people into contact with a large spectrum of other value-seeking people who were willing to invest time and effort in experimenting with co-creation. This is where people like *Nicolette Meeder* and *Marco van Lochem* engaged with the ideology of transformative change represented by the City of Tomorrow and Sustainocracy.
2. It connected value driven initiatives with local city government officials who were often much more developed in awareness and drive for sustainable progress than their counterparts in business sectors. This could be explained by the fact that government relied on tax income, not on banks. Their type of engagement to money and society was different than self-employed individuals who often had the burden of a mortgage, monthly rent and the cost of college-going children.
3. If we wanted to survive, with an aggressive world of transaction and debt-based finance as a point of reference, we needed to develop exactly the opposite (from welfare to wellbeing).

Large business enterprises had both the burden of debt and financial pressure from shareholders. The fragment perception of each societal participant in a money-driven world was based on self-preservation and survival, as shown in this picture (Fig. 2.11).

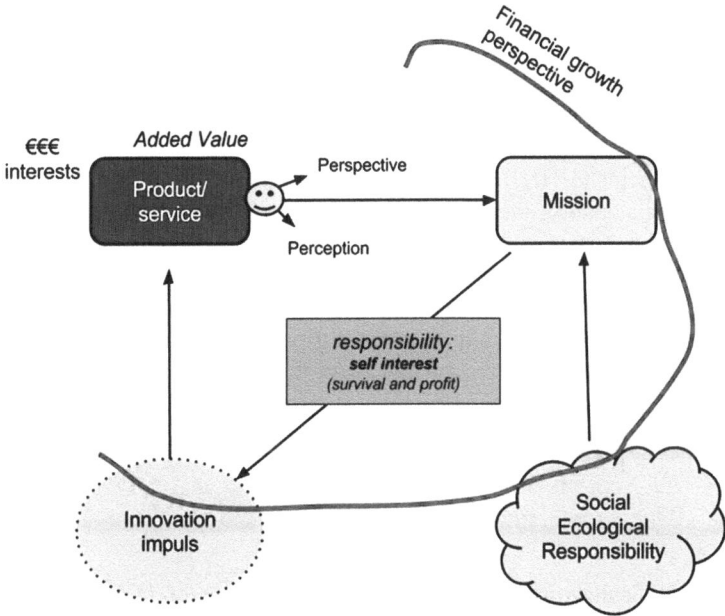

Fig. 2.11 The tunnel vision of fragmented self-interests

City of Tomorrow should not position itself as a new product or service in welfare development, but within the complexity of the holistic wellbeing mission. All the work done in the STIR Foundation had resulted in enormous insights around the transformative complexity that we were facing, but it had not resulted in a definite multidisciplinary commitment or breakthrough. Over 20,000 full-time professional freelance hours had been invested in the learning process. The lack of concrete progression was blamed on the fact that the transitions were nearly always focused on transforming a field of interest that was in the hands of powerful economic and political drivers. Transformative initiatives were always envisaged from within the money-driven culture of economic growth, through product innovation and change of players rather than change of culture. The tension between new ideas and the old establishment was nearly always won by the establishment. In February 2010, it was decided that all the City of Tomorrow workgroups would be dissolved and the ICSE abandoned.

It had become clear that one cannot teach or support others to take on the role of Sustainocrat, because we live in a product- and services-driven culture and structure. If I wanted wellbeing, I had to take responsibility myself, both ideologically and practically. In 2010, the new city council of Eindhoven was addressed in a speech about **Quality of Life**, in which the desire was expressed to co-create a self-sufficient, healthy, energy-neutral city by involving the citizens, government and business innovators. The council reacted positively and asked the representative councilor to discuss the matter with the City of Tomorrow. He waved the suggestion away with the wish to see if the city management could take benefit of the movement by establishing a potential energy and quality of life cooperation itself. There was, again, an obvious economic component to his suggestion, in an attempt to solve the crisis in the city finances.

This political contact and evasive response did not discourage; it just confirmed that fragmented trade issues, such as energy, mobility, housing, education, care, etc., with a powerful economic component and structures of power, would be extremely difficult to tackle using ethics and common sense alone. They would probably need a crisis of their own, a collapse or at least a strong threat before real change would be introduced or accepted. Meanwhile, the battle for ownership and control would continue. In view of STIR, this would further destabilize the community while fragmented self-interests competed for the leftovers of economic drivers in an attempt to sustain themselves at the expense of others. A shakeout was taking place, and everyone was trying to save their institutional souls. Only chaos would lead to openings (Fig. 2.12).

2.1.14   The Amsterdam Internet Congress

The above difficulties in breaking through the status quo of an obsolete but itself sustaining (against all odds) societal structure had become a challenge for STIR. The sum of the competing fragmented interests and reluctance or imposition of each

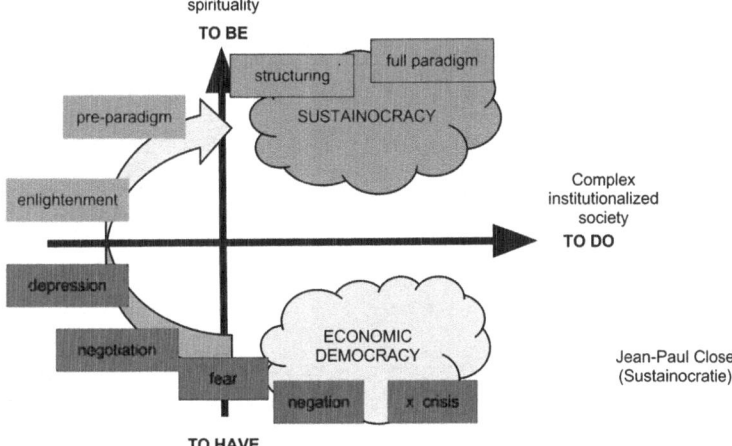

huge spiritual educational process

Fig. 2.12 In 2009 and 2010, negation and fear of collapse was still dominant

fragmented structure to take responsibility for the transformative needs was the theme of a speech during an internet encounter in Amsterdam. Several pioneering organizations from cities around the world participated through video/conference. City of Tomorrow was one of three speakers:

- **City of Tomorrow**: The difficulties in getting fragmented money-dependent structures, including government, to engage in proactive human value-driven responsibilities;
- **Cisco**: Explaining the experiences of the "new way of working" (a type of technology-facilitating home working scenario) in relation to the reduction of CO_2 emissions.
- **University of Madrid**: Showing a 3D presentation of air pollution over the city of Madrid.

Both projects were ending due to lack of funding or the end of a trial period. This experience resulted in the City of Tomorrow combining the above insights with its own STIR initiatives and experiences. Taking air quality, human health and regional dynamics together in a conceptual approach became a source for innovative inspiration for the first time. Cisco did not respond to the invitation to seek continuity of their project in Eindhoven, but the University of Madrid representative did (Photo 2.1).

Owing to the lack of response from Cisco, the STIR Foundation's City of Tomorrow lacked direct access to technology and ICT development structures that could take over the commitment. Within the circle of relationships that had been built up in the City of Tomorrow, Marco van Lochem had signed up.

Photo 2.1 The inspiring image produced by the University of Madrid

- **Marco van Lochem** presented himself as a new age entrepreneur with an extensive CV and relation network in the high tech business world, becoming a sponsor and member of the STIR Foundation. When he heard the suggestion to create a "healthy city" initiative from social and technological perspectives, he decided to commit, and AiREAS was born.

2.1.15 Key Elements that Define "Sustainocratic" AiREAS

AiREAS became the first structure defined primarily from human (not system) ethics and a point of view of responsibility (healthy city mission), with the structure defined subsequently (multidisciplinary co-creation). "To Be" would lead and "To Do" would follow. This turnaround was extremely significant, not just in its positioning as an awareness breakthrough organization but also in its way of working. The key elements were:

- Purpose driven (healthy city)
- Wellbeing not welfare
- No hierarchy (health and air quality dominant, not politics or economics)
- Shared responsibility
- Change driven
- Money not the primary consideration

Fig. 2.13 The AiREAS logo as designed by Marco

This is very meaningful, because it does not simply refer to a business case, it represents a first delicate step into a new type of society defined around a deeper awareness. It would show that evolution is not just limited to individuals but also to the way we interact and create innovative types of communities.

2.1.16   AiREAS

The first working name in the City of Tomorrow was **ID City Home**, referring to the holistic ideology of sustainable progress at the city level. It was the workgroup name in the City of Tomorrow for city development. The name needed to be changed into something that could be related to Air Quality and Regional development. It was Marco who came up with the idea to combine AIR with AREAS. He suggested the "AiREAS" name. The representation of the "i" as a sensor completed the picture, which was also designed by Marco (see Fig. 2.13). The co-creation was operational even if it was only at the level of two individuals.

Meanwhile the AiREAS multidisciplinary venture was being presented to all kinds of potential partners. City officials of Eindhoven had informed us that AiREAS was not necessary because the city was already taking many measures through the Dutch Air Quality Platform that provided funds for infrastructural changes, such as tunnels, traffic light systems, etc. These funds were channeled through the Province of North Brabant. If AiREAS wanted to do something in this field, it would need to address the Province first. This brought AiREAS in touch with two key influencers in the process:

- **Eric van Merrienboer**, former member of the city council of Eindhoven and now director of mobility and economy in the province. Eric influenced the positioning of AiREAS by suggesting the involvement of the four key societal players around the issue: government, science, citizens and business innovators. He was also responsible for the internal distribution within the province of the initial City of Tomorrow proposition to set up AiREAS (Fig. 2.14).
- **Edwin Weijtmans**, air quality program manager in the province. Edwin received the proposition and was very enthusiastic about it. He invited AiREAS (still consisting only of van Lochem and Close) to visit the province, and an alliance grew. Edwin provided the very first small funding (€25,000) for the City of Tomorrow to develop the AiREAS Proof of Concept (Fig. 2.15).

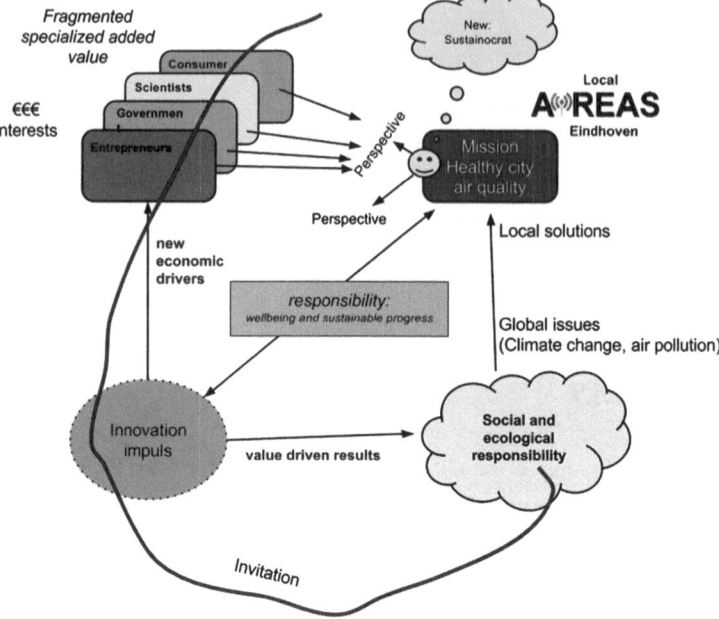

Fig. 2.14 The change of perspective of a Sustainocrat, positioning him/herself in the field of a collective value-driven mission

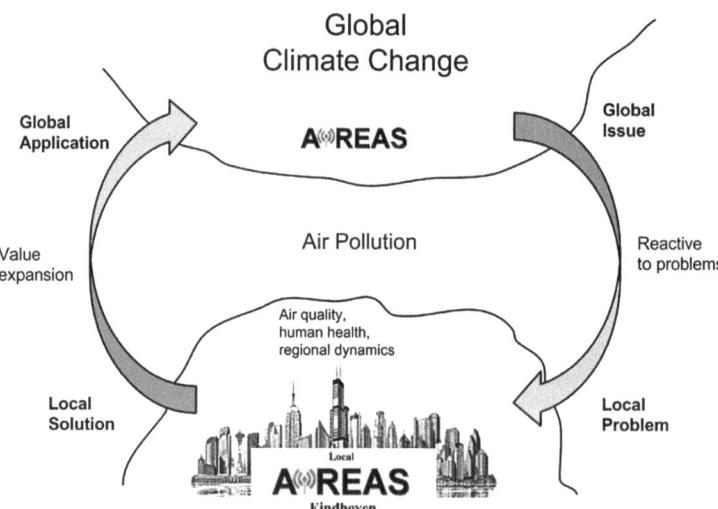

Fig. 2.15 The Local AiREAS mission for human health is inspired by the global issue of environmental pollution and climate change

2.1.17   Commitment First

With the commitment of the province to AiREAS, and the Sustainocratic format to be filled in with multidisciplinary partners, the search continued to develop a consortium. The City of Tomorrow had developed the following value-driven formula to be applied and to distinguish ourselves from the money-driven relationships. Reciprocity is a word that provides a better sense of the return on investment, inhabiting, as it does, a much broader scope than mere financial profit. In the world of trade and financial profits, people engage for just that. In AiREAS, one would engage primarily for "health and environmental quality". AiREAS would be organized in a result-driven, not demand-driven, way, as the mission for a healthy city had been formulated into the very purpose of AiREAS.

Talent $\times$ Input $\times$ Sustainable Human Progress (Purpose) $=$ Innovative steps (Result) $\times$ Reciprocity (Return)

In January 2011, the first true AiREAS multidisciplinary meeting was organized at the Airport of Eindhoven, with the participation of the Intheair.es 3D initiative from the University of Madrid, ITC University of Twente, Philips Lighting, TomTom, Edwin Weijtmans (province of Noord Brabant) and a representative of the City of Eindhoven, together with the two founders. The purpose of the meeting was to cement multidisciplinary support and commitment from the participants for the suggested venture (Fig. 2.16).

All participants but one committed (the one deferring because of internal struggles in the company to define their own purpose for the future). The formal cooperative was then registered. Dutch laws do not yet accommodate the registration of value-driven cooperatives, just money-driven ones. This was the first obstacle, an ideological challenge to the current legislation in Holland which supports just one paradigm (welfare). It was temporarily overcome by incorporating

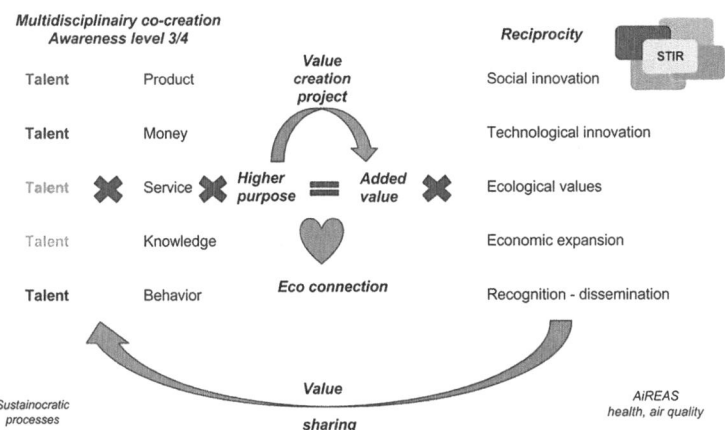

Fig. 2.16 the multidisciplinary sum of talents

constitutional identifiers that allowed us to modify the purpose of the cooperative using our first membership encounter. More obstacles would soon be encountered, demonstrating that transformative change is not just related to practical issues of innovation but also introduces profound discussions on the constitutional and legal formalities that block disruptive processes, standing in the way of sustainable progress. AiREAS was transforming into an initiative that was making the invisible air pollution much more visible.

In June 2011, AiREAS presented its Proof of Concept on a national level in the province of North Brabant. Despite the ideological support of all participants, the invitation to take mutual responsibility was not seconded. The nationwide approach was too far-fetched and the fragmented positions of potential partners too individualized. In some cases, the institutional justification of a potential partner was related to the problem, meaning that elimination of the problem would also eliminate the institution. Institutional self-preservation thus also demanded preservation of the issue, no matter how debatable morally, positioning the organization at the consequence-driven reaction side within the related secondary economy. Comments heard were:

- We don't address over-consumption, we deal with overweight.
- We don't support self-sufficiency because it cannot be taxed.
- We don't deal with healthy air, we repair broken lungs.

This showed yet another complex issue of the dual welfare economy that we had created, the one of economic growth against all ethical awareness, and the economy of speculative care trying to address consequences through remedial tax and insurances. Those in the economic field of growth have an ethical issue to deal with, while those in the economic field of caretaking define their existence in regard to the problem through remedies. Both have a problem with participation in City of Tomorrow transformative processes such as AiREAS because it challenges their own long term existence. Only when the potential partners are aware of the chaos they inflict, or contribute to by their mere existence and management attitude, do they become willing to address their own identity and position themselves experimentally in true value creation. They benefit from it by challenging their own reason to be through the redefinition of their own purpose and contribution to humankind. In this process, many find an unprecedented potential that justifies participation in AiREAS. During the process, it became clear that specialized civil servants were often more advanced in their own entrepreneurship and leadership for regional harmony than business people or scientists. The differentiators in financial and societal engagement of each potential participant became clear, as did the way each perceived the world from their own position. Bridging this perception to the collective, multidisciplinary "healthy city" mission in AiREAS became an awareness trigger for most participants and something they were able to come to terms with personally as well as institutionally. Self-inflicted obstacles, proper to the old paradigm, showed up in the development of the AiREAS commitment, allowing the self-aware institutional executives to redefine their inner structures accordingly.

2.1.18   Territorial Focus

We came to the conclusion as a group that we needed a smaller territory that could act as living lab. This territory should be complex enough to justify a multidisciplinary coalition and co-creative enterprise, yet small enough to produce a single top-down commitment to making it happen by taking responsibility within the connecting process. New city council elections in 2010 had created a new directive in the city of Eindhoven in which applied innovation, civilian participation and sustainability were spear points. The newly installed but experienced city councilor, **Mary-Ann Schreurs**, proved visionary and possessed of a great willingness to participate. "Yes, we want this," was the simple but significant email response that returned the AiREAS effort to Eindhoven. The first Local AiREAS Eindhoven was started as a living lab for applied innovation, citizen involvement and research.

2.1.19   Local AiREAS Eindhoven

In September 2011, the first Local AiREAS Eindhoven meeting was held. I presided over the encounter as a local entrepreneurial civilian representing wellbeing-based sustainable human progress. I had become the first "Sustainocrat." "The world upside down," said the environmental program manager, civil servant **Hans Verhoeven**, who had been selected by councilor Schreurs to represent the local government and its infrastructure. He said this in response to meeting this civilian, who had invited local governance to take responsibility with him "to co-create a healthy city," rather than the other way around. This statement, "the world upside down," became a characterization of the transformative processes in which we had all gotten ourselves involved (Fig. 2.17).

The first multidisciplinary "healthy city with air quality" encounter started with an empty table, no budget, all kinds of possible talented partners, and the higher purpose of a healthy city with air quality and human health as its value-driven purpose. Representatives of government, business, science and civilian groups were assembled around the table. The first priorities and action points were to be established in an open democratic dialogue. The setting was based on equality

Fig. 2.17 AiREAS represents the ideology and co-creative format, Local AiREAS (city) the regional execution

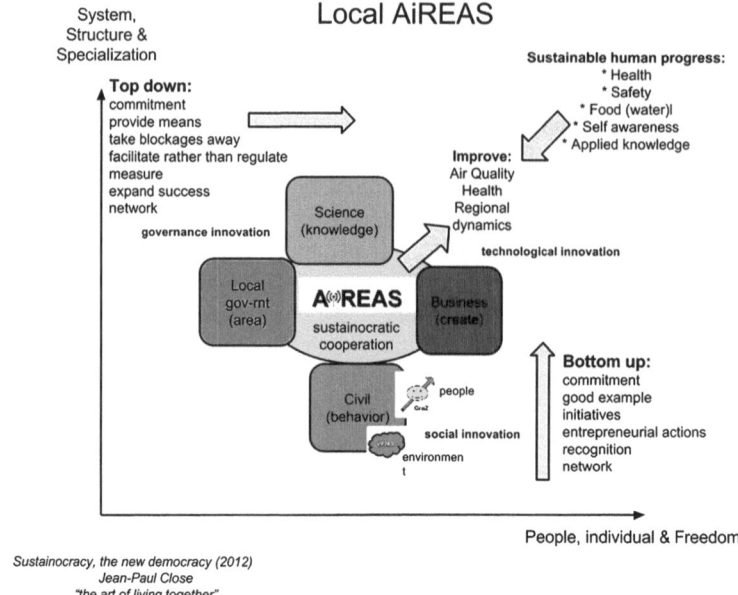

Fig. 2.18 The AiREAS "healthy city" commitment

among all participants. The higher wellbeing purpose of "healthy city through air quality" was what would lead us, not welfare-based politics or economics. No one was "the boss." Each present carried their own responsibility, talent and authority, with the invitation to use it well by creating value together (Fig. 2.18).

2.1.20   First Things First

Councilor Schreurs argued that the city had no own insight into its own air pollution patterns. It depended entirely on the reports that were presented by national organizations such as the Ministry of Health (RIVM). There was no possibility of addressing the issue locally, because the problem was invisible for the local people and policymakers, while interpretation of responsibilities was done outside the scope of the city. The first priority would be to gain insight of our own at the city level. The objective was to gain knowledge about local air pollution in relation to human health. This would trigger value-driven innovation and support important decisions that the city council had to make for the coming decades. We therefore needed to look into the possibility of measuring pollution as closely as possible within the outdoor space of the local population. A first multidisciplinary challenge and priority was born. It became step 1 and phase 1 of Local AiREAS Eindhoven.

2.1.21 Making Visible the Invisible

The project "making visible the invisible" started to take life among the participants, linking the potential of technological and social innovation with ideals of
creating a healthy environment and the need to reflect on city dynamics using real,
locally-validated data rather than those handed down from external authorities. The
need to measure as closely to the population as possible introduced issues like
modelling techniques, presentation and interpretation of data, density of the network so as to be able to arrive at conclusions or cross-referencing and analysis of
data from different sources (e.g., medical statistics with air pollution history), privacy of the population, etc. The scientific partners of different disciplines were
taking responsibility for the application of existing knowledge and the research
potential of the network and mission envisaged. This is described in the Chap. 3 of
this publication. From a technological point of view, choices needed to be made
around available technologies in the market and the purpose that we wanted our
network to serve. These choices are also described in detail in the next chapter of
the book. At this stage, the measurement network was handed down the following
requirements:

- Real time measurements;
- Measure a large spectrum of pollution, including NO_2 and the Ultrafine Particle
  innovation presented by Philips;
- Measure at postal code level (the closest to the human population without
  invading privacy);
- Low maintenance costs and risks (validation, reliability, availability, etc.);
- High quality, calibrated information gathering;
- Low cost (referenced against the expensive official measurement stations used
  by the central government).

A partner consortium, consisting of ECN, Imtech (later Axians) and Philips,
decided to take on the technological challenge with the following distribution of
tasks:

- ECN: equipment design and assembly—Rene Otjes;
- Philips: Ultrafine particles—Ronald Wolff;
- Imtech: data communication and storage—Carl Wolf.

  The team was completed with:

- Scientific insight lung and respiration: IRAS (University of Utrecht)—Dr.
  Gerard Hoek;
- Scientific insight modelling techniques: ITC (University of Twente)—Prof.
  Alfred Stein and Dr. Nicholas Hamm;
- City infrastructure and services: Officials of the City of Eindhoven—Hans
  Verhoeven and Sandra van der Sterren.

The project had highly technological characteristics. It was therefore to be led by co-founder Marco van Lochem as an independent and connecting Sustainocrat. The entire team was given the internal name "ILM" (Innovative Lucht Meetsysteem = Innovative Air Quality Measurement System). Sustainocrat Jean-Paul Close would concentrate on the bigger picture and involvement of the complex "soft side" of AiREAS, the civilian participation, while looking for new steps to take.

2.1.22 *From Idea to Project*

With the abstract mission of a "healthy city" now focused on its first concrete step, the ideological to practical development could be discussed and budgeted. All partners would invest in this first step. Financial means were committed by the two participating governments, the City of Eindhoven and the Province of North Brabant, and the technological partners. The project details were summed up by Marco van Lochem through milestones (Fig. 2.19).

Between September 2011 and January 2012, various meetings resulted in all parties agreeing on the technical aspects of the ILM, its assumptions and expectations, as described in Chap. 3. Despite the financial commitment of the two representatives of government, the project still needed to be passed and approved by all the bureaucratic layers of the city.

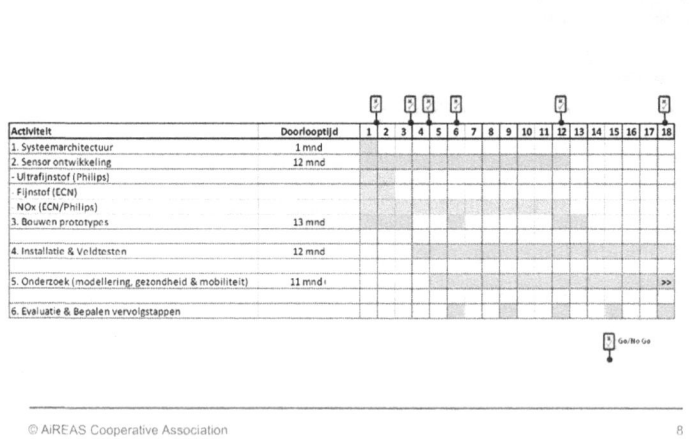

Fig. 2.19 The original planning of the ILM in January 2012

The city of Eindhoven had been affected by the credit crisis and needed to reduce its costs and investment schedules for the coming years. Any new project or financial commitment would require poaching from other plans, commitments or running budgets. All these budgets had already been scrutinized due to the financial savings required. The city was also in the process of drastically reducing its human resources, so little sympathy would be encountered when new projects needed to be accommodated at the expense of others.

The AiREAS community was ready but the participating civil servants now needed time and determination to get the commitment materialized within their own institutions. A global kick-off was announced for June 2012. As this date approached, it became clear that the funding had not yet been cleared by the system. AiREAS decided to go ahead with the meeting, simply to maintain the cohesion of the group and reassure the commitment by all in the process. A new date for the global kick off was set for October 2012. This time, the funds were cleared and phase 1 could start. The October 2012 kick-off became an emotional relief for all involved, a milestone and welcome proof that a complex group could engage in a complex holistic, human value-driven setting. The new paradigm was now a practical reality, and we were going to make it happen. The AiREAS group had taken multidisciplinary responsibility and now had the task of making its commitment come true (Photo 2.2).

Photo 2.2 Marco van Lochem presiding at the October 2012 kickoff

2.1.23   Conclusion About the Coming About of AiREAS

The above story shows how difficult it is to thoroughly engage all components of a complex society in a totally new directive. We can summarize the following aspects as being key to the process:

- Someone, independent of the reigning system, needs to define the common mission and take the initiative of inviting co-creation;
- Depression and (expected) chaos is needed to open the door for propositions of change through awareness and need;
- The right people, with the right need, awareness or mentality and authority, will engage when the proposition suits their interest;
- Key to the territorial partnership is the commitment of local governance;
- The proposition should be complex enough to be challenging beyond the power of influence of the fragmented authority, and small enough to be achievable in a defined time interval;
- At one stage, the initiative should be depersonalized and become a group process with group results;

Recognizing this complex, time-consuming process, it can be repeated as often as global issues demand local solutions. In AiREAS, we have come far, starting with the personal awareness, commitment and determination of a single individual who, after many trials and errors, manages to find and connect people and institutions to make a group commitment. This commitment received the name AiREAS and developed into numerous new age expressions, such as Sustainocracy, multidisciplinary co-creation, the Sustainocrat, value- and result-driven wellbeing-based cooperation, the transformation economy, etc. Without the credit crisis, the doors would probably never have opened to address the issues that we face.

2.1.24   Link with Ethics and Economies

In September 2012, an intellectual gathering in Visegrad, Hungary, with the participation of over 30 countries, discussed the practical evolution of ethics and economies. The presentation and paper about AiREAS in Eindhoven as an evolutionary movement for business and society was accepted and published.[25] It proves the pioneership that AiREAS as a group is introducing for a new interpretation of our reality. The powerful alliance in Eindhoven has proven itself to be a warm, heart-driven commitment that unites seemingly contradictory interests in a common, purpose-driven mission in which the contradictions become complementary forces. The same model can be applied for any issue that humankind faces

---

[25]"The spiritual dimension of business ethics and sustainable management", Corvenius University Budapest 2014—Springer.

in the complexity of sustaining itself as a productive species within the dynamics of an evolutionary natural environment, balancing welfare and wellbeing with wellbeing as the dominant resonance for behavior and structure. Within the current human hierarchies developing capitalism- and consumption-based economics, "ethics" is often defined in legal terms to sustain the political and economic system through lawful public solidarity. In Sustainocracy, however, we place ethics at the level of understanding life and its complex harmony with its surroundings. We develop moral wellbeing around the key values that we have defined for stable communities (To BE). Ethics hence becomes a universal truth of life, not a political or economic one. Having said that, and differentiating now between the transaction-based economy of capitalism and the value driven co-creation of Sustainocratic processes, we can link these systems as well in a coherent self-regulating circular economy (Fig. 2.20).

When we position the fragmented reality of our current institutions, we can conclude that, in our current reality, all operational elements interact within the area of economic growth (the left hand side of the cycle). The necessary self-regulation on the right hand side has not been activated consciously and is left over by the cyclic intervention of nature itself. We become aware through crisis and chaos on the growth side, only then allowing value-driven innovation to initiate a new cycle. By permanently activating the value-driven side, calibrating growth against the harmony of sustainocratic values, a self-regulating mechanism is also introduced into the nature of economics. Without the need to wait for collapse, it challenges growth with continuous change, which is also the way nature works, introducing innovations continuously to sustain life rather than destroying it.

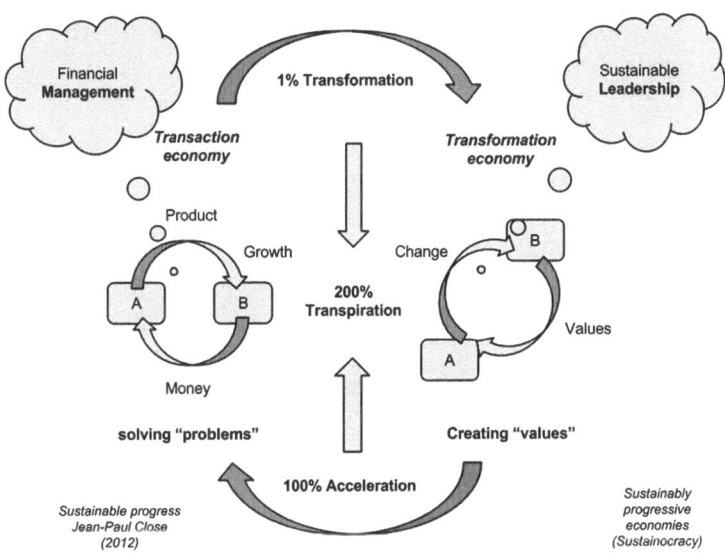

Fig. 2.20 Introducing the transformation economy

This analysis shows how AiREAS is positioned on the side of the transformation economy, resolving harmonization issues with our local health and air quality, while feeding the economy of growth with a whole series of value-driven innovations and proof of new concepts. Ideally, society is permanently positioned on the side of value-driven change while business develops the economy of growth. AiREAS is proof of the principle in this line of thinking.

Chapter 3
"The Invisible Made Visible": Science and Technology

Nicholas A.S. Hamm, Marco van Lochem, Gerard Hoek, René Otjes, Sandra van der Sterren and Hans Verhoeven

An Introduction by Marco van Lochem

As described in Chap. 2, it started for me in 2010. After almost 20 years working in the IT and High Tech Industry, I founded my own company (Odeon Interim Management) and was looking for a way to contribute to a sustainable society. In that period, Jean-Paul Close and I met. Based on his vision and experience regarding sustainability, we discussed how we could improve the living and working environments in cities, initially in The Netherlands, but with a global focus. Polluted air is a major health hazard in world cities and a tremendous cost for society. This was the start of AiREAS, using our network and experience to create a multidisciplinary co-operation with a human value-driven sustainable focus.

In our discussions with the municipality of Eindhoven in North Brabant, the Universities of Utrecht and Twente, ECN, Philips and Axians/Imtech ICT, we defined a first tangible goal and project contributing to the higher AiREAS purpose of healthy cities. We agreed to make visible the invisible by designing and implementing an Innovative Air Measurement Network ('Living Lab') in Eindhoven.

N.A.S. Hamm (✉)
Faculty of Geo-Information Science and Earth Observation (ITC), University of Twente, Enschede, The Netherlands

M. van Lochem
Axians, Eindhoven, The Netherlands

M. van Lochem
Odeon, Best, The Netherlands

G. Hoek
Institute for Risk Assessment Sciences (IRAS), University of Utrecht, Utrecht, The Netherlands

R. Otjes
ECN, Petten, The Netherlands

S. van der Sterren · H. Verhoeven
Department of Environment—Air Quality, Eindhoven, The Netherlands

To get this first project started, the commitment of individual persons from these stakeholders was key (not to mention that it would help in getting commitment from their individual organizations as well). Without this, we could not have been successful. Instead of discussing budgets and investments upfront, we started by co-creating a project plan focusing on 'what has to be done and what are the deliverables.' The next step was to specify the cost of the project. And finally, we asked who would invest and for what would they be paying. It is essential to realize that AiREAS projects are not based on traditional customer-supplier relationships, but on co-creation, mutual commitment and equality.

In this way, we managed to get an agreement on the project plan, including the (fixed) budget and finance part, without losing the entrepreneurial spirit and commitment of individual persons and their organizations. This was very important because of the result-driven characteristic of the project, including the risks. We defined milestones with deliverables and payments and assured everyone that communication and co-operation were open and based on the AiREAS values of 'respect, trust and reciprocity'.

In a relatively short time, this AiREAS co-creation project managed to deliver a world class Air Measurement Network in Eindhoven. And although money and budgets were an important aspect, the focus of participants was mainly on the committed deliverables and contribution to the higher AiREAS purpose. Everybody was aware of the fact that it was a unique initiative (still small, but with huge potential and exposure) and we managed to solve problems and manage risks along the way and within the context of the AiREAS values.

Although it was only the first AiREAS project and new initiatives have already started, with many to follow, it shows that the difference is being made by individual persons taking responsibility. I therefore want to thank everybody involved for their personal commitment to join AiREAS in this great sustainable journey.

Marco van Lochem

3.1 The ILM

This document gives a comprehensive overview of the urban ILM (Innovatief Lucht Meetsysteem, English: Innovative Air Measurement System) that has been installed in the City of Eindhoven under the AiREAS initiative. Here, the intention is to provide the necessary scientific and technical details so that a user can understand the provenance of the data outcome. The social rationale for such a system was outlined in Chap. 2 of this document. Technically, the use of modern, low-cost sensors offers the possibility of obtaining new scientific insights by measuring several air quality variables at a finer temporal and spatial resolution than previously possible. Conventional networks typically measure at only one or two locations in cities the size of Eindhoven, where the temporal resolution tends to be one sample each 24 h (or even coarser).

In brief, the ILM consists of 35 Airboxes which have been installed at various locations throughout Eindhoven. These boxes contain communication and data-logger devices, as well as sensors that measure various air quality variables (particulate matter, ultrafine particle counts, ozone, nitrogen dioxide) and meteorological variables (temperature, relative humidity). These variables are measured every 10 min. Following calibration, these are made available online in near-real time. A complete archive is also made available online.

Particulate matter (specifically PM10 and PM2.5), ozone (O_3) and nitrogen dioxide (NO_2) are the most important air quality variables to be routinely measured. Ultra-fine particles are of increasing interest, but are not routinely measured. Hence, they were included in the set of measured variables. Although the ILM is low cost compared to conventional sensors, there are still cost constraints. The budget allowed for the installation of 35 Airbox sensor units, each measuring PM and O_3. NO_2 is measured at five locations, although there is a plan to expand this to 25 locations (i.e., 20 extra sensors) during 2015. UFPs are measured at six locations.

In order to measure the air quality variables at 35 locations, affordable measurement devices were needed that could easily be located and relocated within an urban setting. As accurate sensors for ambient air were not commercially available, state of the art sensors for PM, NO_2 and O_3 were modified to comply with the required specifications.

In this survey, we first provide an overview of the variables that are measured (Sect. 3.2). The technical equipment and instrumentation are then described (Sect. 3.3), followed by a discussion of data quality (Sect. 3.4). The choice of locations for spatial sampling is discussed in Sect. 3.5, followed by a discussion of data management (Sect. 3.6). Some initial results are presented (Sect. 3.7), followed by a list of projects based on the ILM (Sect. 3.8).

Each section closes with a sub-section labelled "experiences and recommendations." This outlines our experiences to date and gives recommendations for the future. Some of these recommendations are concrete and have been agreed upon. Others recommendations still need to be finalized or further discussed.

3.2 Variables Measured

Table 3.1 shows the air quality and meteorological variables that are measured by sensors in the Airboxes. Further details about the actual instruments are given in Sect. 3.3.

3.3 Instrumentation

This section gives details of the actual instrumentation used.

Table 3.1 Table showing the variables measured by instruments in the Airboxes

Variable	Description	Instrument	No. of locations	Time interval	Units
Particulate matter (PM10)	Particulate matter less than 10 μm (PM10) in diameter	Shinyei PPD42 ECN revised	All	10 min	(Mass per volume) μg m^{-3}
Particulate matter (PM2.5)	Particulate matter less than 2.5 μm (PM2.5) in diameter	Shinyei PPD42 ECN revised	All	10 min	(Mass per volume) μg m^{-3}
Particulate matter (PM1)	Particulate matter less than 1 μm (PM1) in diameter	Shinyei PPD42 ECN revised	All	10 min	(Mass per volume) μg m^{-3}
Ultrafine particles (UFP)/ ultrafijnstof	Particle number concentration	Aerasense NanoMonitor PNMT 1000	6	10 min	(Particle count per volume) # cc^{-3}
Ozone (O$_3$)	Ozone concentration	E2V MICS 2610	All	10 min	(Mass per volume) μg m^{-3}
Nitrogen dioxide (NO$_2$)	Nitrogen dioxide concentration	Citytech Sensoric NO$_2$ 3E50 ECN revised	5[a]	10 min	(Mass per volume) μg m^{-3}
Temperature	Air temperature	Sensirion SHT75	All	10 min	Degrees centigrade
Relative humidity	Relative humidity	Sensirion SHT75	All	10 min	Percentage
Date/time	Recorded as UTC/GMT. May need to be adjusted to CET/CEST for communication	SIMCom SIM908	All	10 min	Uses unix time with time zone UTC/GMT
Coordinates	GPS coordinates, longitude, latitude, altitude	SIMCom SIM908	All	10 min	Degrees, minutes seconds

[a]20 extra sensors will be added during 2015, bringing the total to 25 sensors

3.3.1   The Airbox

The Airbox was developed to serve as weatherproof housing for an array of sensors. On the lower side, well ventilated space with 3 grates is reserved for mounted sensors. A 1 mm gauze is applied to prevent insects and large particles from entering. The lockable box (brand Sarel) is made of Polyester with outer dimensions of 43 × 33 × 20 cm and designed to be attached to street light poles. It carries a battery as its power supply. The battery is recharged daily during nighttime hours. The Airbox is 12 kg and 5 W.

Photo 3.1 Airbox

The UFP sensor (AeroSense Nanomonitor) is located in a separate box. This UFP box (30 × 20 × 17 cm), also by Sarel, is made of ABS/PC and attaches easily to each Airbox (plug and play). The UFP box is supplied with its own battery, 4 W and 8 kg.

Both boxes are mounted onto street light poles, the Airbox at a height of 2.5–3 m and the UFP box between 2 and 2.5 m (an example is shown in Photo 3.1).

Both boxes are CE—EMC (Conformité Européenne—Electromagnetic Compatibility) tested and approved.

The Airbox has several interfaces that communicate with the sensors and the modem. An overview is given below.

- GPRS GSM interface for transmission of sensor data and download of firmware files;
- 10-bit and 24-bit analogue interfaces for the measurement of the battery voltage, PM sensor, ozone and NO_2 sensor;
- SPI interface for temporal data storage on a SD-card;
- I2C interface measurement of the micro controller print card temperature and storage of parameters;
- RS232 interface for debugging information;
- JTAG programmable interface for the microcontroller.

The microcontroller is the basic centre of the Airbox. It samples all sensors, does certain calculations and sends the accumulated data by GPRS and through an Imtech/Axians server towards an application on the ECN server. This application permanently saves the raw data in a database. In case of server or GPRS network outage, the accumulated data is saved on the Airbox SD-card. When the server and GPRS network is resumed, data not yet transferred is automatically sent afterwards.

3.3.2   PM (PM10, PM2.5, PM1) Sensor

The basic sensor is the Shinyei PPD42, revised by ECN for improved performance. The optical sensor consists of an IR LED and a photo-transistor detector. Flow and drying of the particles is established by an electric resistor in the sensor container. In addition, the dark current of the cell is retrieved. Results are averaged over 10 min and transmitted to the ECN server. PM10, PM2.5 and PM1 concentrations are calculated sequentially.

3.3.3   UFP Sensor

The *NanoMonitor* is a small, wall-mountable device for detecting ultrafine particles in the 10–300 nm size range. The functionality of the NanoMonitor relies on electrical charging of particles in a sampled airflow and a subsequent measurement of the particle-bound charge concentration. The sensor signal is an electrical current measured by a sensitive current meter and represents the particle charge captured per unit time in a Faraday cage. The current is derived from the total charge on all airborne particles obtained after their charging in a high-voltage corona section. To reduce signal drifts over the course of time, the device periodically performs an automatic zero-offset check (typically once every 5 min).

The NanoMonitor has its own box and can easily be attached to the Airbox and moved to another according to the plug and play concept.

3.3.4   Ozone Sensor

Ozone is measured by the E2V MICS 2610, a MO_x (metal oxide) sensor that changes conductivity characteristics through ozone adsorption. The sensor is locally heated to 350 °C, but also corrected for variations in ambient temperature. In the Airbox, three ozone sensors are implemented in order to enhance the precision and reliability of the operation. The sensors are, on a monthly basis, verified by monitors operated by the national air quality network.

3.3.5   NO₂ Sensor

The NO_2 sensor is based on the electrochemical cell Sensoric NO_2 3E50 by CityTech. In order to make the sensor applicable for ambient air, it was revised to deal with interferences by trace gasses and water vapor. A differential measurement set up with a switching valve and reagent cartridges was established in front of the detecting cell. Concentration calculations take place on the ECN server. NO_2 measurement resolution is 10 min.

3.3.6 Temperature Sensor and Relative Humidity Sensor

Temperature and RH in the AirBox sensor compartment are measured with a Sensirion SHT75. The SHT75 is a digital pin-type humidity and temperature sensor. A capacitive sensor element is used for measuring relative humidity while temperature is measured by a band-gap sensor. Due to instrumental heat generation, the temperature in the Airbox is on average 3 °C higher than the ambient air and appropriate corrections are subsequently made. T and RH can only be used for indicative purposes.

3.3.7 Electromagnetic Compatibility (EMC)

In order to obtain the CE approval, an Airbox equipped with UFP was tested (by Dare) according to the EMC (Electromagnetic Compatibility) directive. The following tests were performed:

- Conducted emission test (class A) according to EN55011 (2009) + A1 (2010)
- Radiated emission test according to the same standards

  The above tests judge the EMC effect on other equipment.
  The following immunity tests were performed:

- Harmonics according to EN61000-3-2 (2006) + A1 (2009) + A2 (2009)
- Flicker according to EN61000-3-3 (2008)
- Voltage dips and interruptions EN61000-4-11 (2004)

  These tests judge the effect of external equipment on the Airbox and UFP.
  The following tests were conducted on the Airbox controller print:

- Power supply checked with calibrated electrometer. Voltage deviation should not exceed 10 %
- SD-card checked on partition type and volume, as well as initialization ability
- Modem communication while located outdoors

  Finally, the battery was tested. This showed that the battery can supply sufficient power for at least 18 h.

3.3.8   Experiences and Recommendations

Section 3.7 outlines certain results that show the functioning of the sensors within the Airbox and ILM system. However, it is too early to evaluate the long term performance of individual sensors, the Airbox or the ILM. This will be evaluated technically during the interim calibration (see Sect. 3.4.2) and at the end of the expected life of the project (5 years). It will also be evaluated through user experience.

3.4   Data Quality

There are three components to the evaluation of data quality.

(1) Regular **calibration** is the formal calibration of the ILM instruments against standard instruments, as well as the inter-calibration of the ILM instruments. This is further subdivided into initial calibration, interim calibration and preventative maintenance.
(2) **Validation** is the process of checking the data to ensure that they adhere to predefined quality standards.
(3) **Smart spatial data quality** evaluation and **online normalization** are research topics concerning the development of novel methods for low-cost sensor networks.

To date, only the initial calibration (part of 1) had been finalized and implemented whereas (2) is in progress (3) will form part of the DAMAST research project and may be a component of other future research projects.

It is important that the outcome of any data quality evaluation [whether (1), (2) or (3)] is routinely reported with the data. This is discussed in Sect. 3.6 (data management).

3.4.1   Regular Calibration and Preventative Maintenance

This section describes the basic calibration of the instruments whereby the sensors are calibrated against recognized reference instruments.

3.4.1.1   Initial Calibration

The first the set of AiREAS Airboxes were operated outdoors for extended periods of time at an ECN test site. In this phase, all sensors were compared to the median time series of each sensor type (PM, O_3, etc.) for this period. This way, for each sensor, the deviation in terms of offset and slope was calculated in comparison to the median time series.

Secondly, the correlation coefficient for each sensor (expressed as R^2) with the median time series was derived. This value was used as a criterion for proper operation of a specific sensor. In case this criterion was not met, the sensor was rejected. The criteria for PM, O_3, T, RH and UFP were 0.8, 0.9, 0.8, 0.8 and 0.95, respectively. The test was based on comparison constraints retrieved by simultaneous, co-located, outdoor operation with at a minimum of 10 Airboxes for at least three days. All sensors (PM, Ozone, UFP, RH and T) were inter-compared and normalized to the median values based on the 10-min aggregated values. The average R^2 was as follows: PM 0.89, O_3 0.97, T 0.92, RH 0.98 and UFP 0.98. These all meet the above-mentioned criteria.

Next, a subset of three Airboxes was calibrated against reference equipment at an urban background site for two weeks. The reference equipment for PM10 and PM2.5 was the Met One BAM (Beta Attenuation Monitoring) and for ozone (UV photometry Thermo).

The UFP (Nanomonitor, Aerosense) was calibrated against the GRIMM SMPS (L-DMA, CPC5410) at the ECN site. The T and RH sensors that are inside the Airbox were considered indicative.

NO_2 sensors were introduced into the AirBoxes later on. In 2015, NO_2 sensors were added to five Airboxes with a plan to introduce a further 20 sensors later in 2015. These NO_2 sensors will be calibrated against a reference NO_x monitor (chemoluminescense) in Eindhoven.

3.4.1.2 Experiences and Recommendations

A problem with the implementation of the ILM has been the lack of planning and budgeting for data quality evaluation. This is an important lesson to be learned as we go forward with the ILM and as similar networks are rolled out in other cities. Recommendations for data quality evaluation for the remainder of the ILM lifetime are set out below. These should be evaluated at the end of the ILM lifetime.

It should be noted that the data quality evaluation in low-cost sensor networks is an important research topic. Traditional approaches to calibration and validation tend to be costly. A low-cost network needs a smart, low-cost data quality evaluation protocol.

3.4.1.3 Interim Calibration

After a fixed interval, each Airbox will be removed and the sensors will be calibrated against appropriate reference sensors.

The intention is to calibrate all sensors on a regular basis with certified reference equipment. The sensors are calibrated as contained in the Airbox to avoid artifacts possibly introduced by the housing. The sensors are compared with the reference equipment while exposed to ambient air for at least 48 sequential hours. The location is by preference within the application area, in this case, in the city of Eindhoven. Reference equipment is operated under certified conditions following the issued EU directives. The RIVM LML stations are suitable for this purpose. Proposed calibration frequency is once per two years. Calibration results are used to assess sensor performance characteristics and might lead to an adjusted calibration frequency. Interim calibration of Airboxes is carried out in batches (typically 1/3 of the total number of Airboxes per batch) in order to minimize disturbance of measurement series.

UFP sensors (NanoMonitor, Aerosense) cannot be calibrated in this way, as this parameter is not measured by the LML. Therefore the UFP's will be calibrated by the manufacturer. The intended calibration frequency is once a year.

Interim calibrations should, where possible, be coincident with preventive maintenance. At such an event, calibration will be executed before and after the maintenance to cover for possible induced changes in sensor performance.

3.4.1.4 Preventative Maintenance

Individual sensors (either the whole device or individual components) have a limited lifespan. Hence, a preventative maintenance program is required. The maintenance program can be combined with interim calibrations of all other sensors, including those newly replaced.

Preventative maintenance is scheduled on a biannual basis. This frequency is based on the manufacturer information of the lifespan sensitive components in the Airbox. The most important parts to be replaced are the electrochemical NO_2 cell, the NO_2 sensor cartridges, the O_3 sensor, and the Airbox and UFP box battery. Furthermore, the PM sensor will be cleaned.

After a final functional check, the serviced Airboxes with sensors will be calibrated as set out in the previous chapter. In cases of preventive maintenance coinciding with interim calibration, the Airboxes will also be calibrated beforehand.

3.4.2   Validation

This has two forms: online and afterwards. The objective is to evaluate whether the data are valid in the sense that they match what we expect from the calibration. This is less stringent than calibration, but may identify, for example, drifts in the calibration or gross errors. Possible outcomes are (i) do nothing (the data show no problems), (ii) apply adjustments/corrections, or (iii) interim calibration. The methodology for online and afterwards validation will be based on the procedures applied by RIVM (Ministry of Health) to the LML (Landelijk Meetnet Luchtkwaliteit), the nationwide measurement network.

3.4.2.1 Online

The online validation concept consists of two steps. First is the automatic check based on the internal sensor diagnosis. This is sensor-dependent. For example the dark current measurement on the PM sensor is monitored. The next step is the test on the data consistency. Outliers are invalidated. Sudden jumps in sensitivity, negative and out-of-range values, as well as flat line evolution are detected and invalidated. Corresponding criteria are managed in the metadata database. Finally, non-sensor specific information is taken into account. For example, if an Airbox is in service, the measures will be invalidated.

Furthermore, various other processes are monitored in the Airbox, such as battery voltage, processor temperature, and modem and SD card characteristics.

3.4.2.2 Afterwards

A monthly check will be performed by a validation operator. During this manual operation, sensor values of one AirBox are compared to other neighbor stations, as well as being checked for consistency within that one Airbox. The step is important because not all error values can be detected automatically by software. Also online invalidated values are reconsidered. Furthermore in case interim calibrations have been performed the correct implementation is considered and sensor with an abnormal behavior invalidated.

3.4.3 Smart Spatial Data Quality Evaluation and Online Normalization

This has been left as a research topic. Indeed, spatial data quality is an explicit part of the DAMAST project. The idea is to develop lightweight methods that can be used for data quality evaluation, validation and online normalization. A low cost sensor network requires lightweight validation. This is an active topic of scientific research in which Hamm and Stein are active.[1] There is already much research in

---

[1]Zhang, Y., **N. A. S. Hamm**, N. Meratnia, **A. Stein**, M. van de Voort and P. J. M. Havinga (2012). "Statistics-based outlier detection for wireless sensor networks." International Journal of Geographical Information Science **26**(8): 1373–1392.

the context of air quality—but mainly for data having a coarse resolution in space and time. The challenge is to develop approaches for the fine temporal and spatial resolution data that the ILM delivers.

3.5   Locations and Spatial Sampling

The choice of locations for the Airboxes was the subject of extensive discussion. After developing a general set of criteria, Sandra van der Sterren (Municipality Eindhoven) prepared a selection of sites with pictures to judge the suitability. This selection was evaluated by the AiREAS team, particularly by representatives of IRAS-UU, ITC-UT and ECN. After several iterations, a final selection of 35 sites was made.

The main goal of the network is eventually to map air quality and its change over time. This means that we need to understand the link between spatial and temporal variability and local sources. In practice, this may require the consideration of different scales. This begins with generic sources (roads and industry), as well as particular sources (traffic lights, roundabouts, building works, airport). Locations where people are potentially exposed are also of interest. This includes variation within the areas where people live and work. The whole city should be addressed, not just the centre; the whole population, including the most vulnerable. Finally, the new network should be coherent with the existing network of passive NO_2 measurements.

The following starting points were identified to select the sites:

– The main criterion was to select monitoring locations that are relevant for human exposure of residents of Eindhoven. All measurements were thus at sites relevant for representing exposures near homes, schools or other buildings. For example, we did not select sites at a major roundabout that may present high concentrations, but would not represent residential exposure. We further selected sites on major streets that represented residential exposure and thus tried to avoid measuring directly at the edge of a road. We also avoided such areas as industrial sites.
– The second key criterion was that the Airboxes needed to be attached to lamp posts (which provide electricity). This clearly limited the choice of locations.
– Sampling heights were a compromise between safety (not easily reached by third persons) and the desire to represent exposures. Sampling heights between 1.5 and 4 m have often been used in previous networks and research studies. In AiREAS, all Airboxes are mounted at a height roughly between 2.5 and 3 m, the UFP boxes between 2 and 2.5 m. Especially in busy streets (close to a source), this may modestly underestimate concentrations for traffic participants.
– Measurements sites should cover background locations and busy roads in about equal proportions. Busy roads were overrepresented compared to their occurrence, because they will likely be an important source of spatial variation of air pollution.

- Measurements (especially on busy roads) should be taken at a distance from the road (i.e., not directly at the curbside).
- Measurement locations in neighbourhoods should be spread over the whole city, including some neighbourhoods on the outskirts of the city. These are quiet but also close to the motorway.
- Measurements are often made at houses and schools on busy roads (e.g., on the ring road, the inner ring road and other important roads). On these roads, measurements should be performed at their representative parts. If a significant portion of a road runs through a canyon, measurements should be performed in the canyon and not on a smaller, more open portion of the road.
- Some measurements were made at locations where there are known complaints or citizen concerns.
- Measurements should be made at the same locations as instruments from the existing municipal NO_2 network or from RIVM (Rijksinstituut for Volksgezondheid and Milieu, National Institute for Public Health and the Environment).
- Offices and hospitals are less relevant than schools and homes because they tend to have circulation systems, meaning that they are less sensitive to air quality.

Information on the current Airbox locations is given in Table 3.2. Along with details of the locations, this also states which sensors have the UFP and NO_2 sensors (Photo 3.2).

3.5.1 Experiences and Recommendations

At this point, it is too early to evaluate the choice of Airbox locations. We only have one full year of data and not all sensors were installed (e.g., the NO_2 sensors) or properly calibrated from the start (e.g., there were initial problems with the O_3 sensors). We expect to be able to comment further on this after the second year.

Over the course of 2015, a further 20 NO_2 sensors will be added to the network, bringing the total number of NO_2 sensors to 25. There will still only be six UFP sensors, which is why the rotation scheme (Sect. 3.5.1.1) is proposed.

3.5.1.1 UFP Rotation Scheme

UFP is only measured at six sites. Two are urban background sites and four sites are located on busy roads. The limitation to six sites arose due to the relatively high cost of the UFP sensor. This limits the information that can be obtained about the spatial distribution of UFPs throughout the city. For this reason, we intend to implement a rotation scheme in which UFP sensors are moved between locations.

Table 3.2 Locations of the airboxes at the time of first installation

Location	Postal code	X	Y	Address	Airbox no.	NO₂	UFP no.	Location-description
1	5627 TE	158608	389159	Finisterelaan 45	19	x		Edge of residentialarea, nearby A2/A50 high density traffic [about 104 m between Airbox and driving lane (wall and houses in between)]
2	5626 BN	157966	388001	Amstelstraat	17			Residentialstreet near elderly home
3	5629 NK	161214	389171	Falstaff 8	5	x		Residentialstreet, area near A2/A50 (measurement is for back ground information)
4	5628 PZ	161106	387853	Maaseikstraat 7	9	x		Along residentialstreet near a park
5	5632 DN	162548	387177	Grote Beerlaan 15	6	x		Along residentialstreet
6	5622 HV	160177	386018	Rijckwaertstraat 6	28			Along residentialstreet
7	5612 NJ	160502	384307	Lijmbeekstraat 190	2			Along residentialstreet
8	5613 EE	162421	383403	Sperwerlaan 4A	20			Along residentialstreet
9	5652 SN	158275	383875	v. Vollenhovenstraat	24			Along residentialstreet in neighborhood between speedway and motorway
10	5657 AR	157305	383585	Sliffertsestraat 12	13	x		Child care at edge of new residentialarea near A2/N2 high traffic (distance 400 m)
11	5655 JJ	158078	380509	Twickel 30	14	x		Edge of residentialarea, near A2/N2 De Hogt high density cross road [distance to N2 is 78 m, to A2 is 110 m (buildings in between)]
12	5654 DT	160191	381309	Jan Hollanderstraat 70	12	x		On residentialstreet

(continued)

Table 3.2 (continued)

Location	Postal code	X	Y	Address	Airbox no.	NO$_2$	UFP no.	Location-description
13	5644 HL	161702	380451	Vesaliuslaan 50	32			On residentialstreet
14	5611 HV	161675	383158	Spijndhof	30	x	3	Small square/parking area in city center with little traffic
15	5614 EP	162461	382142	St. Adrianusstraat 30	27			Along residential street near high traffic ring
16	5646 JM	163804	380995	Eij-erven 41	1			Along residential street at the edge of town, nearby small park
17	5641 PX	164504	383907	Donk 24	21	x	2	Along residential street at edge of town, near small park and school
18	5612 EJ	161212	384829	Pastoriestraat 57	34	x		Primary school De Driestam, along busy crossroad. Pastoriestraat/OL Vrouwestraat 25,000–35,000 vehicles per 24 h day average
19	X	159849	382317	v. Weberstraat-Limburglaan	4	x		Secondary schools along Ring; Christiaan Huygens College and St. Lucas. Limburglaan 32,000 vehicles average per 24 h day
20	5623 NR	161815	385212	Hudsonlaan 694 (Kennedylaan)	36	x	5	Residential apartment building along Kennedylaan/very busy road
21	5621 JC	160154	385070	Boschdijk 393	35	x		Houses along busy road near 393
22	5651 LZ	159010	383907	Noord-Brabantlaan 36	39	x		Houses along busy road, near the Evoluon, nearby RIVM-national measurement station
23	5616 JG	159478	383152	Botenlaan 135	7	x		Houses at busy road/ring
24		160818	382949	Mauritsstraat bij gemeentelijk meetstation	25	x	1	Houses along busy road/West tangent

(continued)

Table 3.2 (continued)

Location	Postal code	X	Y	Address	Airbox no.	NO$_2$	UFP no.	Location-description
25	5611 GD	161328	383077	Keizersgracht 28	3	x		Houses along busy road/inner ring
26	5611 DM	161588	383286	Vestdijk bij Pullman-hotel/Gedempte gracht 109	26	x		Houses along busy road/inner ring
27	5613 GC	163160	383161 (is weg)	Jeroen Boschlaan 170	23	x	4	Houses along busy road/ring
28	5644 PA	162234	381613	Leostraat 17	11	x		Houses along busy road/ring
29	5643 AJ	162517	381356	Leenderweg 259	8	x		Houses along busy road. Leenderweg outside the ring, high traffic density
30		160769	383032	Mauritsstraat bij Anna v Egmondstraat	40			Close to airbox 24
31	5504 GD	163429	384034	Hofstraat 161	22	x	6	Quite road near rail track (Eindhoven-Helmond)
32	5625 EA	160917	386622	Genovevalaan	37	x		National measurement station RIVM opposite shopping center Woensel along fairly busy road
33	5582 EJ	ca 160267	ca 378349	Vincent Cleerdinlaan, Waalre	31	x		Houses in/nearby wood bos in Waalre, quiet region, light post at end of road/beginning bicycle path
34	5651 CD	159254	384161	Beukenlaan 62	16	x		Office buildings along a busy road/ring
35	5631 BN	161896	385000	Ds. Fliednerstraat	29			Maxima Medical Centre (Eindhoven)/Hospital at certain distance from busy road/Kennedylaan

A full spreadsheet can be found as an annex to the section
X and Y indicate the easting and northing (in metres) direction according to the Rijksdriehoek (RDH) coordinate reference system. RDH is the coordinate reference system for the Netherlands

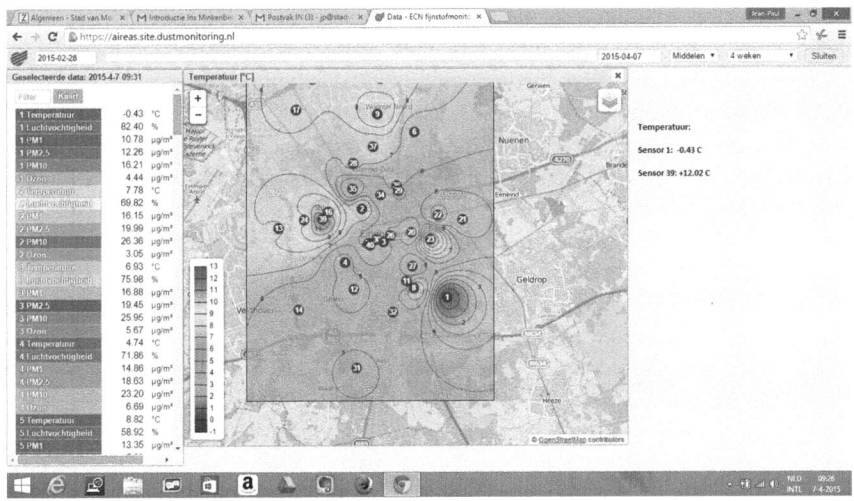

Photo 3.2 Picture of Map of Eindhoven with all ILM points

When combined with correlations with PM, NO_2 and ozone, we will then be able to build a model for the spatial distribution of UFP throughout the city. Such rotation schemes have been applied in other studies.[2,3] The rotation cycle will be completed after one year. We will then evaluate the results to determine whether the rotation scheme should be changed. This evaluation will be based on principles of spatial statistical analysis—including modelling, mapping and sampling design.[4,5,6]

[2] Eeftens, M., M. Y. Tsai, C. Ampe, B. Anwander, R. Beelen, T. Bellander, G. Cesaroni, M. Cirach, J. Cyrys, K. de Hoogh, A. De Nazelle, F. de Vocht, C. Declercq, A. Dedele, K. Eriksen, C. Galassi, R. Grazuleviciene, G. Grivas, J. Heinrich, B. Hoffmann, M. Iakovides, A. Ineichen, K. Katsouyanni, M. Korek, U. Kramer, T. Kuhlbusch, T. Lanki, C. Madsen, K. Meliefste, A. Molter, G. Mosler, M. Nieuwenhuijsen, M. Oldenwening, A. Pennanen, N. Probst-Hensch, U. Quass, O. Raaschou-Nielsen, A. Ranzi, E. Stephanou, D. Sugiri, O. Udvardy, E. Vaskoevi, G. Weinmayr, B. Brunekreef and **G. Hoek** (2012). "Spatial variation of PM2.5, PM10, PM2.5 absorbance and PM coarse concentrations between and within 20 European study areas and the relationship with NO_2—Results of the ESCAPE project." Atmospheric Environment **62**: 303–317.

[3] **Hoek, G.**, K. Meliefste, J. Cyrys, M. Lewne, T. Bellander, M. Brauer, P. Fischer, U. Gehring, J. Heinrich, P. van Vliet and B. Brunekreef (2002). "Spatial variability of fine particle concentrations in three European areas." Atmospheric Environment **36**(25): 4077–4088.

[4] **Hamm, N. A. S.**, A. O. Finley, M. Schaap and **A. Stein** (2015). "A spatially varying coefficient model for mapping PM10 air quality at the European scale." Atmospheric Environment **102**: 393–405.

[5] **Stein, A**. and C. Ettema (2003). "An overview of spatial sampling procedures and experimental design of spatial studies for ecosystem comparisons." Agriculture Ecosystems and Environment **94** (1): 31–47.

[6] **Stein, A.** (1997). Sampling and efficient data use for characterizing polluted areas. In V. Barnett and K.F. Turkman (eds) Statistics of the Environment 3—Pollution assessment and control. Chester, Wiley.

The selected rotation scheme involves keeping one UFP sensor at a fixed location for a full year. The other five UFP sensors are kept in one location for 3.5 weeks and then moved to the next group of five locations. Thus, in 25 weeks, all 35 locations of the network can be measured once. The cycle is then repeated, meaning each site is measured for two 3.5 week periods during the year. We choose two periods in the year to avoid making comparisons between, for example, summer measurements in one group and winter measurements in another. Although rotation means that the average concentration of a site does not formally represent a true annual average, previous work has shown that, after adjustment for temporal variation, measured at a continuous reference site, spatial differences between sites can well be represented.[7]

The fixed site should be an urban background location, that will be used to correct the measurements at the other five sites for differences in time, following procedures in previous research studies (see footnote 7).[8] Each group of five being measured simultaneously should ideally represent a diversity of sites, that is, busy streets and background locations; city centre and suburban sites in different neighborhoods.

3.6 Data Management

An efficient and effective data management protocol is essential for various reasons:

- the data need to be retrieved and archived in a reliable fashion;
- various processing steps are necessary before the data can be made available to the user. These processes need to be tracked and executed;
- raw and processed data need to be archived;
- metadata need to be made available to the various users. This metadata should include the data quality information.

The main data flow is illustrated in Fig. 3.1.

The raw data is generated locally in the Airbox. The data is sent every 10 min to Axians by GPRS. Axians passes the data through to ECN. ECN performs the calculation, validation and metadata management. Metadata comes from calibration and other services. The processed data are then communicated back to Axians who make it public.

These steps are explained in more detail below.

---

[7]Diamond, J. (2011) *Collapse: How Societies Choose to Fail or Succeed.* Penguin Books; Revised edition.

[8]Other STIR initiatives to date are: FRE2SH (eco-city: local self-sufficiency and productivity), STIR Academy (educational triple "i" platform: inspiration, innovation, implementation) and SAFE (safety and social innovation).

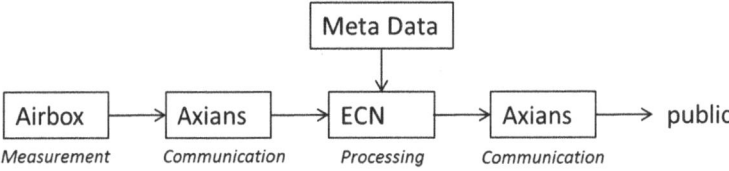

Fig. 3.1 The main data flow

3.6.1 The Airbox

At the Airbox, raw data are collected from each sensor by the microcontroller (Atmel AT90CAN128). This is done by means of 10-bit and 24-bit ADCs. In the processor, all signals are processed and averaged. (Plans are also in the works to calculate the noise level of each sensor.) A data string of 73 defined data fields is created every 10 min. Through a SPI interface, data strings are temporarily saved on an SD card. The data remains saved on the SD card until it is sent through GPRS GSM to Axians.

3.6.2 Axians (1)

Axians receives the raw data from the Airboxes and checks for a correct format. The raw data is saved in an HDF5 format. Then, this data set is forwarded directly to ECN.

3.6.3 ECN

The process is illustrated in Fig. 3.2. The Airbox data coming through Axians is collected by an Internet server and saved in a database. A direct communication line with the Airboxes makes firmware updates possible. The ECN server saves the raw data in the database.

External data is collected and saved into the database on a continuous basis. This includes, for example, information coming from the LML (Landelijk Meetnet Luchtkwaliteit (Dutch national air quality monitoring network)) stations in the region of Eindhoven.

The incoming data saved in the database are processed continuously. The raw 10-min values from the sensors are converted into concentration values using conversion formulae and constants maintained in the metadata database. The calibration parameters per sensor, also coming from the meta-database, are then applied. The processed data are then saved in the database and forwarded to Axians.

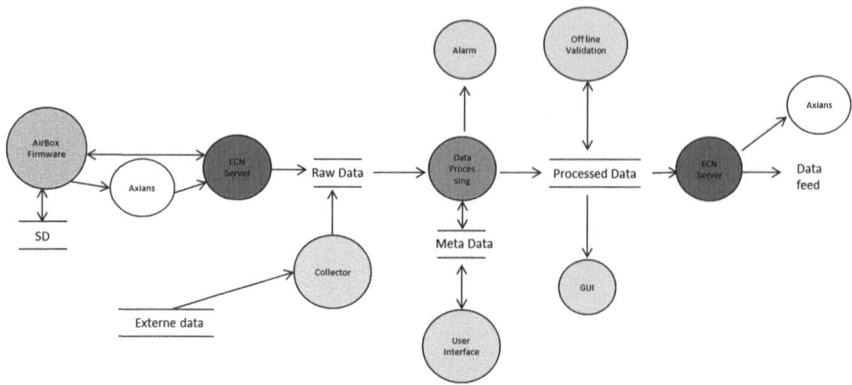

Fig. 3.2 The process at ECN

3.6.4 *Axians (2)*

The data are made available in three formats:

1. HDF5 (hierarchical data format version 5) files for each day. HDF is a self-describing data system which can store both data and metadata. In principle, this could contain metadata about the sensors (i.e., the lineage of the data), units and data quality information. This format (and a similar format, NetCDF) is widely used for archiving and serving environmental datasets. For example, it is used by NASA to archive and serve remotely sensed imagery. This was the rationale for Imtech (contact Carl Wolff (Axians)) to adopt this data format, and the data archive was initially only available in this format. Unfortunately, no metadata have been provided and the HDF file contains a series of tables containing the data from each sensor. A further problem in working with these data is that they do not correspond to a strict 24-h period. The data can be downloaded from: http://82.201.127.232:8080/ (accessed 28/6/15).
2. CSV (comma separated value) files for each sensor. These CSV files contain the complete dataset for each sensor since the sensor was installed (predominantly November 1, 2013). The CSV file is updated daily so that it is never more than 24 h old. The CSV files correspond to individual tables provided in each HDF file, except that they are for ALL days, not just the previous 24 h. This method for serving the archived data was introduced in autumn 2014 as an alternative to HDF. Although HDF is potentially richer in the sense that it allows more information to be archived, it has not been used to its full potential. Given the data that are provided, CSV works equally well and is more straightforward for certain users. Ease-of-use is the rationale for making the data available in this format. The data can be downloaded from: http://82.201.127.232:8080/csv/ (accessed 28/6/15).

3. Finally, the most recent data are made available in real time. For example, the most recent measurements for Airbox 1 are made available at http://82.201.127. 232:3011/api?airboxid=1.cal[9] (accessed 28/6/15). The rationale for this approach is that the data are made available in real time. Using some basic software tools, a user can download these data and manipulate them in his or her own software.

3.6.5 *Experiences and Recommendations*

The data management and data access procedures are described above. It is highly positive that the data are freely available, although they have mainly been used by Axians (formerly Imtech), the ITC-UT and by Andre van der Wiel (Scapeler).[10] The data are content-rich and valuable from both a scientific and societal perspective. Unfortunately, various problems have been encountered when working with these data, including:

(1) The data are not easy to access. In particular, the HDF data are not easy to work with.
(2) There is a lack of metadata. This includes basic things, like the time zone.
(3) The individual tables are inconsistent. For example, the tables for Airboxes 26 and 35 have different column names than the other tables. The ordering of the tables is also different. This means that anybody wanting to work with these data must first spend time solving what should be a simple database design problem.
(4) There are several incidences of missing data.
(5) Some Airboxes have been moved since the installation of the network. Some have later been put back.
(6) At some point, there was a switch from recording floating point numbers to recording integers. According to ECN, this is because this is the limit of the precision of the instruments.

In future, the system for archiving and serving the data should address the following points.

(1) The individual tables should be consistent.
(2) Metadata should be made available with the data. This should include a basic description of the sensor, the units, time zone, etc. In the long term, the data quality information (including data quality flags) should also be provided. This should be thorough and complete.

---

[9]The IP address may change due to structural changes in partner relationships.
[10]Scapeler—www.scapeler.com.

(3) The data should be archived and served in a more robust and user-friendly way.

(4) We should look for alternative formats to serve the data. One suggestion has been XML (which allows the values and the metadata to be provided). An alternative could be an appropriate open-source database (e.g., PostgreSQL), together with a sensor observation service (SOS). An SOS provides an interface that allows data to be accessed directly from software over the Internet. The eventual solution will be discussed and agreed upon with the primary users.

In the future, ECN plans the following activities, which will link data management to the work on data quality.

In the coming year, an online validation procedure and an alarm function will be added to the process (see section "Online"). A further plan is to add an "afterwards" (see section "Afterwards") validation process, according to the RIVM LML validation strategy. Here, a skilled operator manually checks the dataset on a monthly basis and makes a final decision as to whether the processed values are valid or not. All data entries will be accompanied by a flag indicating the quality of the value. Based on this information, the value can be treated as fit-for-purpose or not. A GUI (Graphical User Interface) will offer users the possibility to look at all historical data in various ways.

A user-friendly interface would make input to the metadata database possible according to strict formats. The interface will also make it possible to perform queries and disclose metadata according to a user-defined structure. For now, this is a labour intensive activity.

Currently, the processed data are forwarded directly to Axians. In future, data processed according to the online validation strategy will be transmitted to Axians for display purposes only. The definitive data, validated according to the "afterwards" protocol, will be made available online.

3.7   Results

This section outlines some initial results. These link mainly to the developmental and calibration activities and to data quality checks.

3.7.1   Initial Tests of Sensors

Initially, in the summer of 2013, the Airbox sensors were tested under operational conditions. This was accomplished by comparing an individual sensor's measurement values with the average of the total set of sensors. Also, the relative sensitivity

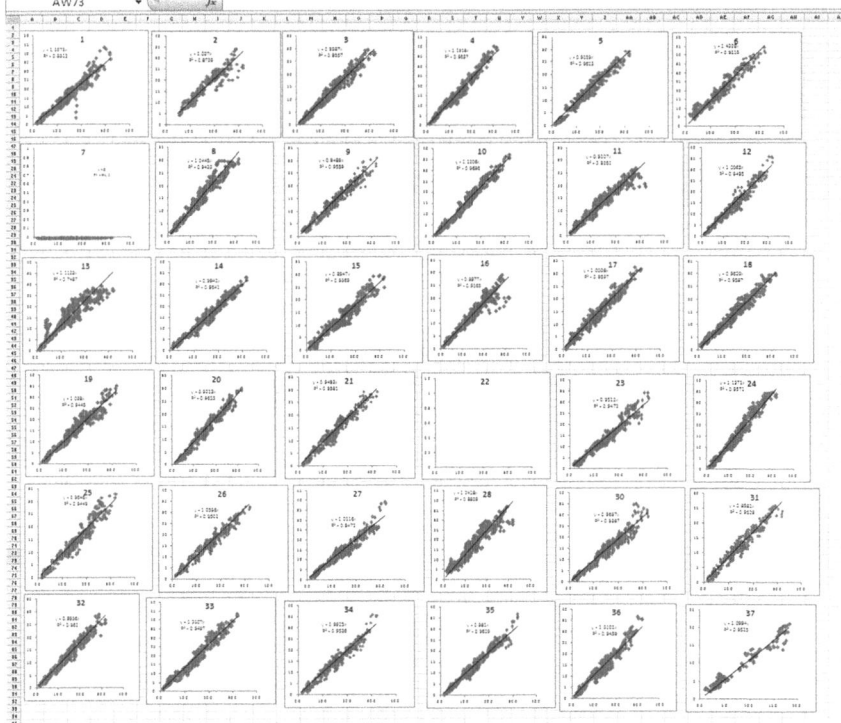

Fig. 3.3 Intercomparison PM measured per-sensor

of each sensor was determined. Figure 3.3 shows an example of the inter-comparison of one of the PM channels.

In November 2013, airboxes were operated sequentially at a reference site. The Airbox sensors for PM and ozone were calibrated against certified instrumentation. Examples are shown in Fig. 3.4.

3.7.2 Evaluation of Sensor Precision

In order to evaluate the precision of the whole sensor network, we undertook analysis during episodes of stormy weather. During such an event, the sensors are exposed to well-mixed air, and the hypothesis is that the air quality should be similar in different locations across the city. Although local effects may still be present, they will be small (relative to calm weather conditions), due to the high dilution effect. All sensors are expected to measure similar concentrations. Figure 3.5 shows an example of PM2.5 concentrations during a storm event on October 28–29, 2013.

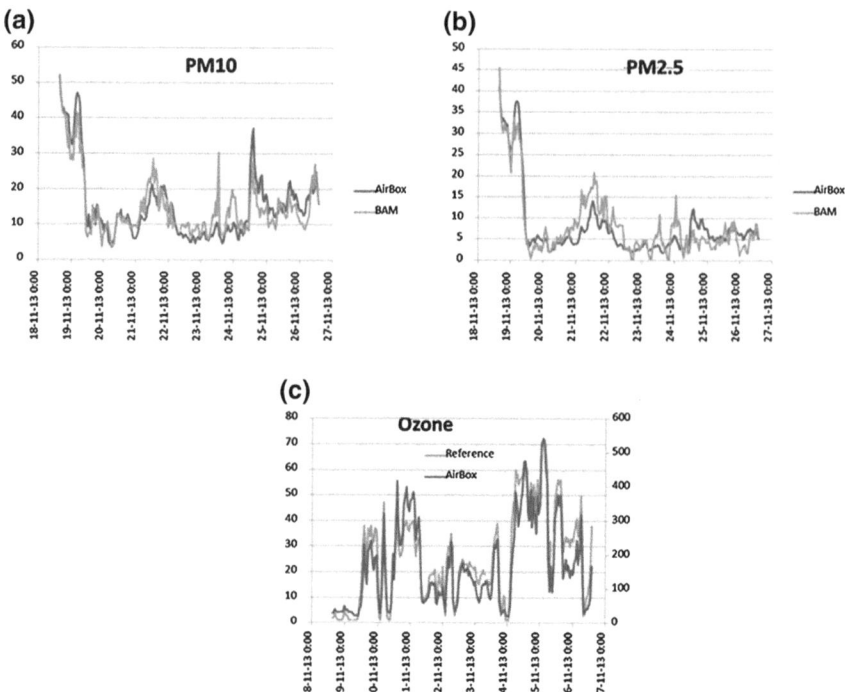

Fig. 3.4 Comparison of airbox measurements to reference measurements

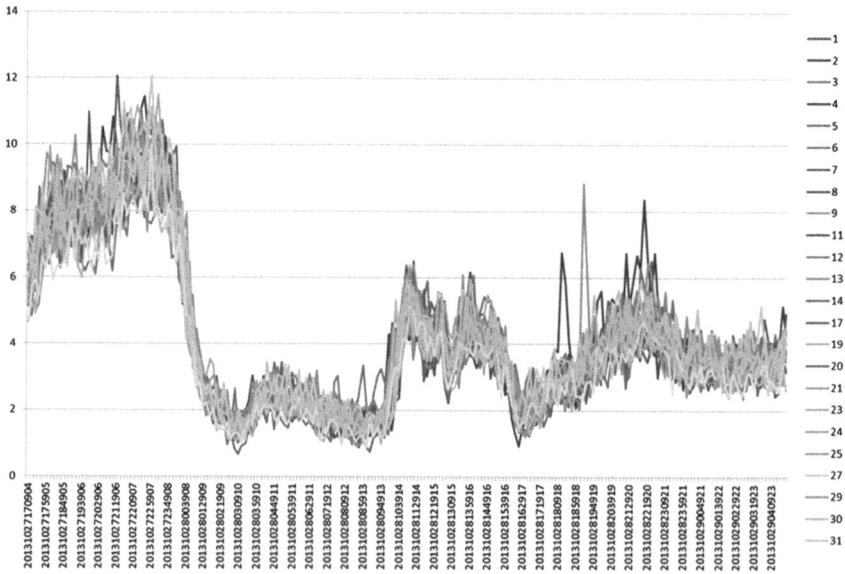

Fig. 3.5 Temporal profile of PM2.5 measurements for all sensors during the storm event of 28–29 October 2013. Units of concentration: $\mu g\ m^{-3}$

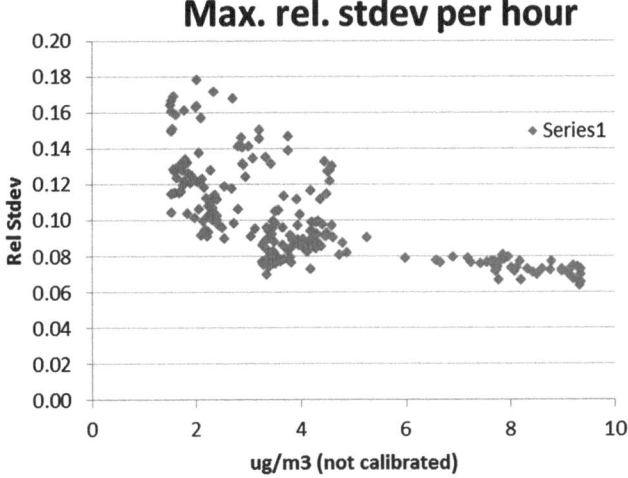

Fig. 3.6 Relative standard deviation for PM2.5 during the storm event of October 28–29, 2013

Comparing the relative standard deviation as a function of the measured concentration reveals a good precision of better than 8 % for concentrations higher than 6 μg m^{-3} (see Fig. 3.6).

The NO$_2$ sensors were installed in autumn 2014. A set of four sensors were co-located and evaluated at an urban background site in Eindhoven (Mauritsstraat, 2014) for a period of one week. Figure 3.7 shows the deviation of an individual sensor against the median of the others. Figure 3.8 shows the relative standard deviation of the four sensors. The relative standard deviation is of the order of 15 %, although it is higher at very low concentrations.

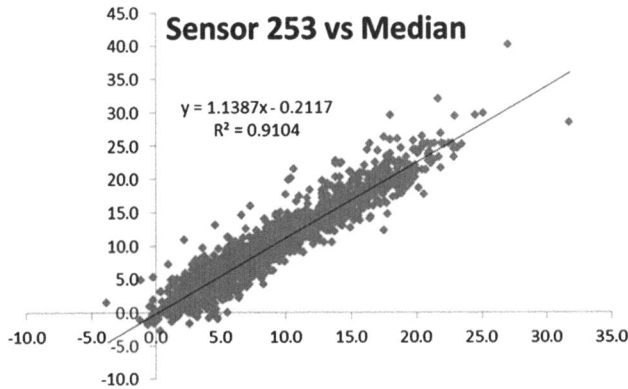

Fig. 3.7 Deviation of an individual sensor against the median of the others. Units of concentration: μg m^{-3}

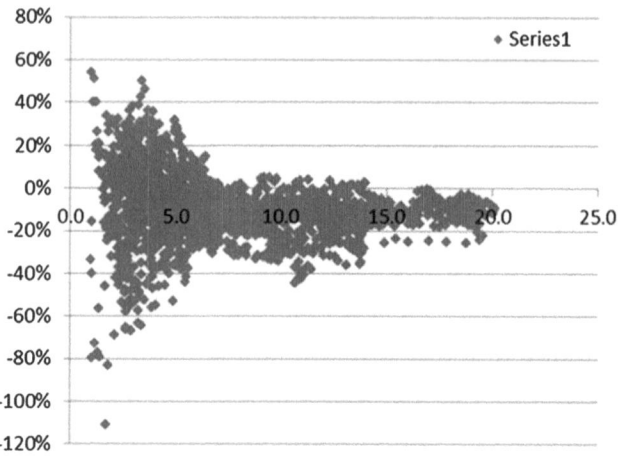

Fig. 3.8 Relative standard deviation of the four NO$_2$ sensors

3.8 Scientific Projects Based on the ILM

Since 2013, various projects of different size and duration have been developed based on the ILM, These all contribute to the AiREAS ideals. These are listed below.

(1) B.Sc. project of H. van Gurp

- van Gurp, H. 2014. Spatial data quality of air quality data collected at the city level. Determining the spatial data quality of the provided by the AiREAS project at the municipality of Eindhoven. Report for B.Sc. minor project, Faculty of Geo-Information Science and Earth Observation (ITC), University of Twente.
- van Gurp was one of the first users of the ILM data served by Axians (Imtech at the time). He provided useful comments on the usability of the HDF data, as well as insight into the reliability of the early ILM data, particularly the O$_3$ data.

(2) STW (Dutch Technological Foundation) Maps4Society call awarded the project Development of an Automatic system for Mapping Air quality risks in Space and Time (DAMAST) to ITC-UT and IRAS-UU. This will fund a promovendus (doctoral candidate) and the associated research. The doctoral candidate began on 1 September 2015.

(3) M.Sc. project of Lingyue Kong

- City-level air pollution modelling and mapping

(4) M.Sc. project of Edgardo Alfredo Vasquez Gomez (Alfredo)

- Service-based sharing and geostatistical processing of sensor data to support decision-making
- Alfredo's thesis provides valuable insight that will help with the development of a data management framework for DAMAST and for AiREAS more generally. Alfredo is working at ITC-UT for the second half of 2015, before returning to a position in his home country of Guatemala.

Chapter 4
Experiences After 5 Years of AiREAS and 1 Year of ILM

Jean-Paul Close, Sandra van der Sterren, Marco van Lochem, René Otjes and Mary-Ann Schreurs

4.1 The Way Things Work at AiREAS

The process of figuring out the way things would work at AiREAS had been completed in one initial loop, referred to as the **STIR loop**. This means that, from an empty table and a shared higher purpose, a wellbeing-based, human values-structured project produced measureable results that could be expanded across the world through welfare-based economies. The loop added unique new values to the community. These values enhance the potential sustainable human progression through steps towards better air quality and health while each has an economic potential on the world market through expansion. A new economy and economic model arises and proves itself upon closure of this loop. It became an example for the world of how trade- and growth-oriented structures could engage in wellness based commitment trusting that elements would appear that enhanced their global competitive positioning. This is what makes this exercise so unique and interesting, much more than the simple design of a technological measurement system (Fig. 4.1).

During the AiREAS general members meeting of January 2014, this working model was the one that elicited the most praise and came to be considered one of the key values of AiREAS for expansion worldwide. By that time, the members and participants had had a lot of experience with the model. This has been captured in

J.-P. Close (✉)
STIR Foundation, Eindhoven, The Netherlands

S. van der Sterren · M.-A. Schreurs
Department of Environment—Air Quality, Eindhoven, The Netherlands

M. van Lochem
Axians, Eindhoven, The Netherlands

R. Otjes
ECN, Petten, The Netherlands

© The Author(s) 2016
J.-P. Close (ed.), *AiREAS: Sustainocracy for a Healthy City*,
SpringerBriefs on Case Studies of Sustainable Development,
DOI 10.1007/978-3-319-26940-5_4

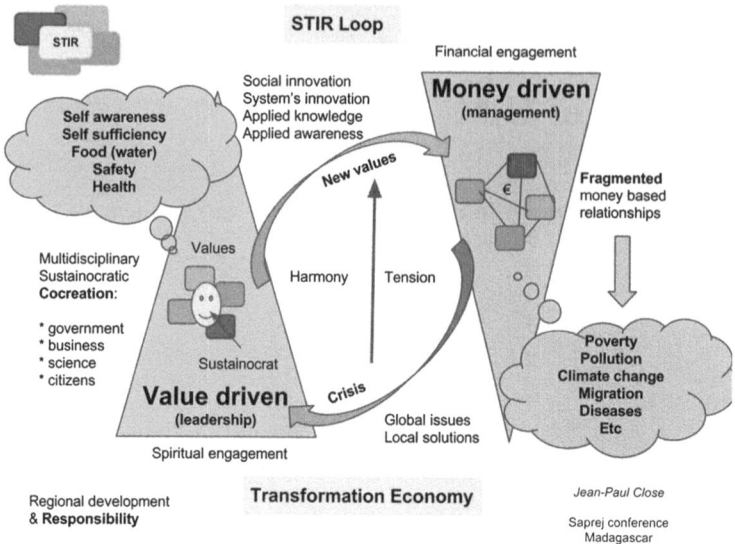

Fig. 4.1 The STIR loop starts with awareness when tension becomes too large (Money driven = welfare- and Value-driven = Wellbeing)

the STIR Academy of the STIR Foundation for educational purposes and implementation worldwide.

However, one loop around alone does not win a battle for the human species, as human complexities continue to challenge our harmony with ourselves and our environment. The fact that the loop now exists and has proven itself means that it can be applied as often as necessary to reduce our vulnerability while benefitting from both the values created and the new economic cycles it announces.

Key in starting the STIR loop is the generally felt need for change, the measurable "tension" that provokes awareness development and the start of the loop. Instead of waiting for the chaos that can collapse financial, societal and even biological systems, one accepts the invitation to introduce multidisciplinary change. STIR always uses core natural human values to define the issue and the higher purpose. In the case of AiREAS, this was clearly defined by "regional air quality, human health and regional dynamics." The reason why partners are attracted and relate to the issue is up to them, and so is the diversity of reciprocity obtained by participating.

By defining and accepting the key human values that need to be protected and enhanced to assure a harmonic society and an ecological relationship, the process for deciding to address the issues when they are in jeopardy is no longer a democratic one. It becomes a leadership issue that can be executed at any time, and when necessary, without having to wait for general elections or budget rounds. The working procedure of AiREAS applies when awareness unites the right disciplines (Fig. 4.2).

Fig. 4.2 The AiREAS
workflow from freedom to
structured project processes

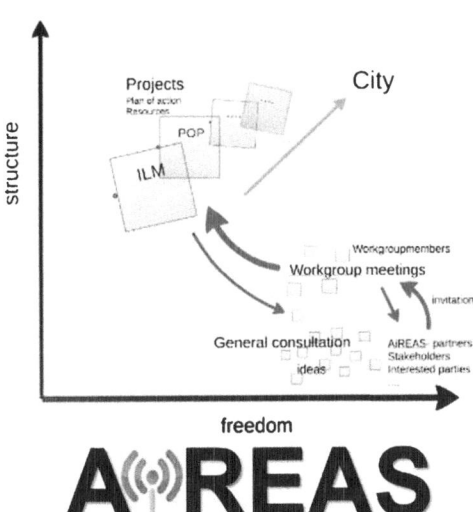

Workingprocedure

4.1.1   The Workflow in AiREAS

The workflow, once the loop is started and captured in a name (s.a. AiREAS), has three steps:

Stet 1:   *Relating democratic freedom to the higher purpose*

The leading motivation in AiREAS is the higher purpose of co-creating a healthy city, using air quality, human health and regional dynamics as points of measurable reference. Anyone, member or non-member of the cooperation, can propose initiatives that contribute to the mission. In the case of the ILM development, for instance, the suggestion came from the city's councilor. The idea to set up AiREAS in the first place had come from a civilian. Twice a year, AiREAS gets together to simply gather and interpret ideas. When the AiREAS group accepts that the idea, or the suggestion, is promising enough for the mission, the next step is initiated. A simple rule for step 1 applies to all members:

> Whatever you can do alone, you do alone. Whatever part of the mission is too complex, requiring the involvement of the others, is done together.

This simple rule avoids potential misuse of the co-creative capacity of AiREAS or the development of competing interests between the members and the group. Initiatives taken on by the AiREAS group are therefore always compliant to the higher purpose of local wellbeing development and the multidisciplinary, result-driven process of the group. Every step that is completed, such as the availability of the ILM, provides a new set of instruments for the next steps, in which any new organization can be involved. Every new step always starts with an empty table again. The loops that were completed generate trust in the process and proof of principle to those who were not convinced enough to participate in the first rounds but will do so for the next.

Step 2:   *Workgroups are formed*

The idea or co-creation proposition now needs to be worked out into a project. A workgroup is formed with those people, partners, members and institutions that wish to participate, protecting at all times the balanced configuration of a sustainocratic venture (government, business, civilians and scientists). During the workgroup encounters, the project is worked out in detail with all the elements needed, including the individually talented or specialized involvement of each participant, their particular responsibility, the expected results of the project and the financial requirements. Decisions are made on each aspect, including the impact on the city, the involvement and stimulus of the city's population, and the source and structuring of the funding.

During this phase, there is complex negotiation so as to make everything fit, from practicalities to formalizing commitments. The latter is a challenge on its own, because the co-creative participative effort of each of the partners is done from their own perspective of self-interest, often strongly coloured by their speculative welfare origins. The commitment needs to be tied to backing from their individual institutions. Even if a manager, a top executive or key politician is enthusiastic about the AiREAS mission in which he/she got engaged, the backing of the institution they represent needs to be engaged as well. In practice, during the encounters, we first relate to the human being at the table and the commitment to human values from a personal perspective. Only after establishing the importance of the steps to be taken does the professional contribution and authority become relevant.

In many cases, the institution is governed by the old fragmented principle of economy of growth (business and scientific research centers) or political confinements (city government), not the immediate backing for value-driven change for measureable wellbeing that expects an investment in time and talent. The economy of growth (welfare) argument is overcome by the promise to present the project involvement as a potential authentic driver of innovation, with new global growth potential through the traditional transaction economy. Not many potential partners understand this while focusing on short term survival and risk avoidance. They see AiREAS first as a potential customer to sell products to, instead of a societal R&D in which unique values are co-created, tested and proven for the world market using their products and innovative capacity. Those who do understand create a self-selecting process between the interest in participating and the final formalization of the commitment.

A simple formula is applied here:

Membership is free of charge, not free of commitment.

In every new AiREAS project, the entire process of choice and commitment starts again. The fact that we have gone through the process already once before is a positive reference for subsequent cycles. Once committed to the results of the workgroup, the third step can begin. Partners repeatedly commit time after time, project upon project, determining through self-leadership how far they want to go. Without projects, AiREAS ceases to exist, showing that it is not just a self-sustaining initiative, but a value-driven movement. AiREAS depends on the need to create local healthy environments and measureable wellbeing through the willingness of partners to take responsibility together by defining projects. The driving force is often the sustainocrat who maintains the focus on the higher purpose and develops a group's cohesion by introducing challenging encounters.

AiREAS has no resources of its own other than its mission, its bonding way of working and the quality of the commitment of the participating partners. No contracts are involved; just the strength of a result-driven purpose and trust, making the venture unique in the world.

Stet 3: *Project execution*

When everything is clear, funding and commitments confirmed, and expected results defined, the team is ready to bring the project into execution. Steps 1 and 2 require interaction at the executive level where responsibility can be taken directly in committing to a process. Step 3 can be delegated to personnel of the participating institutions.

The mix of people involved is unique and interesting. We see civilians participating free of charge out of personal interest for a healthy living environment. Or they develop entrepreneurial initiatives around the wellbeing mission which they test in the AiREAS network. There are self-employed professionals linking large institutions with fragmented specializations to the project's complexity through the budgeted platform. And we see well-paid professionals from big institutions bringing in their expertise and a large company's potential. Civil servants facilitate the activities in the city and often defend the use of public tax money as part of the financial commitment. All the participants have their own uniquely different reciprocity expectations in the project and still complement each other effectively in the value-driven process.

> Reciprocity is not just expressed in money, but also in the field of knowledge development, measurable healthy city and personal health development, worldwide product and concept expansion potential, political and social recognition, validation, participation, celebration, team ownership, visibility, etc.

4.1.2   *Financial Routine in AiREAS*

We come from a world structure in which everything is predefined in financial terms *before* initiating a process. This has proven to be a highly ineffective way of working, creating a lot of unnecessary bureaucracy and fragmented interests, consuming debt before values are created, if they are created at all. The process of

AiREAS (and any other sustainocratic venture) is exactly opposite. We start with an empty table, without any means, just an abstract, holistic higher purpose for creating human wellness, with partners and whatever means that may be available from society all involved in a result-driven process. No one in AiREAS is paid to be present or to participate. Everyone is invited to trust one's own potential, talent and reciprocal interests in the process.

This is an extremely difficult process in which to initiate people. Many talented self-employed people, for instance, cannot spend two years in a value-driven process with the risk of "no go" without the allocation of compensating funds. They are mostly in short term survival mode in the still dominant world of welfare and trade, and so need to commit only to part time or even wait until the process is close to being completed before agreeing to a "go." The local self-employed are not interested in global expansion and expect their contribution to be expressed in direct local reciprocity. Bigger organizations do have the breadth but find themselves emerged in the market-driven pressure of volume and short term results, often in crisis-managed reorganizations. A middle way is to try to define projects that are complex enough for multidisciplinary co-creation while small enough for a faster throughput. Or we can define steps in between as predefined milestones. For the local contributions a special reciprocal value system can be considered.

This value-driven commitment, therefore, has tended towards a self-selecting nature of participating talents and institutions. The consequence of working with a higher purpose based on a global humanitarian or environmental issue is that any innovative idea is welcome. People educated in the field of budgeted financial economics tend to feel submerged in a process they don't understand when entering AiREAS. Everything seems to happen at the same time, requiring each participant to experience awareness breakthroughs. Everyone undergoes a learning curve. Interestingly, a lot of projects can appear and develop at the same time in all kinds of fields related to health, city dynamics or air quality. And every project has its own unique composition of participants.

4.1.3   Confidence Based Interaction

How fast can a complex, multidisciplinary project be organized? That all depends on the level of awareness, commitment and confidence in the participating members. The ILM was extremely complex, and required intense scientific, technological and political interaction in a time of financial crisis and organizational uncertainties for all corporate members.

To keep the group together, delicate interaction was needed that regularly reconfirmed commitment all the way up to the final allocation of the funds for step 3 to take off. This process has also delivered a lot of new insight into and knowledge about value- and result-driven processes in multidisciplinary human complexity. The mix of human and institutional behaviour in a string theory environment of commitment became a field of experience of its own that not

Fig. 4.3 AiREAS partners
commit to this column of
values

Universal values
Ecological values
Human values
Economic values
Value driven cooperation
Trust
Respect
Equality
Individual talent/authority

Column of values
Sustainocracy - Close 2012

everyone understood, especially newcomers to the groups. This was also captured
in the City of Tomorrow's STIR Academy for expansion into the world.

The column of values defined in the City of Tomorrow, and referred to by Marco
van Lochem in his introductory note, was put into practice in AiREAS. It reads as
follows (Fig. 4.3).

The financial backing of the project was not process-driven but rather
result-driven. The investment of the first "Sustainocrats" (Jean-Paul and Marco)
was their own in time and effort for the start-up years. This was necessary to assure
their independence from money-driven control mechanisms and decision-making
and their ability to steer processes out of the old paradigm. They had to try to
sustain themselves in the old money-driven reality and find time to coordinate the
value-driven processes through steps 1 and 2 up to step 3 of AiREAS. This was not
easy, but determination and trust in the mission made it worth their while. Initiation
of step 3 can hence be seen as a milestone for the sustainocrat, while it is an
operational kick-off for the partners.

The financial structure is therefore as follows:

Steps 1 and 2:   AiREAS uses the infrastructure and facilities of the participating
partners. No costs are involved for AiREAS while usage of space
and catering is seen as representation costs for the partner. No one
receives any payment of fees in this entire process.

Step 3:   Means are allocated, including actual financial means, as opposed
to all the resources that partners may provide (buildings, infrastructure, personnel, etc.) which are investments as well. Money is a
means, just like talent, authority, commitment, technology,
knowledge, etc. They all are an investment in the concrete value
creation processes defined in the project.

For the sustainocrat, the start of a project is a milestone that is rewarded through
a percentage in the financials of the project. For the other participants, it is a kick-off

to co-create a new set of values with financial backing for their efforts. The following financial formula is used for 100 % of the financial commitment (real money allocated to the project).

100 %	Project value < 500.000 euro (%)	Project value > 500.000 euro (%)
AiREAS overhead (sustainocrats)	10	5
Education, group network support (new Local AiREAS) and representation	10	5
Operational capital (result driven)	80	90

There is always a natural grey area between the switchover moment of <=> 500 K€. This is dealt with transparently within the dynamics of each project, and with the participants settling somewhere between 10 and 5 %. The same formula applies when the operational capital is divided over operational groups that have their own overhead which is managed by self-employed individuals who act as sustainocrats in their subgroup, linking their activities with the others. This way of working is not meant as a hierarchy but as uniformity in equality within the operational processes. Within the allocation of operational capital, differences may apply because of the participation of all kinds of organizations, each of whom have their own operational reality. Sometimes we see a sustainocrat who takes on certain operational tasks too. This is done when the experimental phase requires the effort and knowledge they can provide or when no other professional can be found to do the job.

"Result-driven project operations" means that partners are not charging input based on hours and material invested but on expected and measurable outcome. Partners are expected to do their part in the commitment. The values that are created have worldwide potential but only when the results have been finalized and made visible. For the business partners, the economic profit is not in the co-creation itself, which can be seen as a societal multidisciplinary R&D, but in the expansion of what has been created together. It is, hence, an investment. Since local government and citizens are the direct beneficiaries of the co-creation, it is logical that they participate in the labor and financial backing. But they cannot be treated like a cash cow. Equality remains important and money cannot be dominant, a position always reserved for the results, measured against the higher purpose. It is the task of the sustainocrats to keep that framework centered among all the participants.

4.1.4   October 2013 General AiREAS Participants Meeting

With the availability of the ILM in September, the finalization of phase 1 is coming into sight. A general partner and participant meeting was called for to determine the

Table 4.1 New potential projects

Number	Proposed by	Proposition	Group decision
1	Dr. Eric de Groot	Research project of 4000 local citizens on health in relation to air quality	Yes
2	Ben Nas	Involve city quarter FRE2SH activities (another City of Tomorrow initiative) in AiREAS	Yes (maybe combined with 1)
3	Marco van Lochem	Co-create an App for mobile phones to show air pollution status in real time	Yes
4	Nicolette Meeder	Investigate behavioral issues (criminality, mental health, etc.) in relation to air pollution	No (maybe combined with TU/e research project)
5	Marco van Lochem	Integrate ILM with traffic management system	Maybe if positioned as health co-creation, not commercial
6	Doctor's post	Establish an AiREAS research office in the new city center development of Strijp-S together with doctor's post	No, due to absence of proposer
7	STIR Academy	Entrepreneurial push around ILM	No time to address this proposition

next steps for AiREAS in the healthy city project. Seven suggestions were proposed by members (Table 4.1).

At the time of finalization of this analysis, several of the proposed actions have reached project status, showing that the STIR loop is being continued. At the same time, new challenges have been introduced into the finalization of the first phase, ones that we deal with in this manuscript. We needed to ask ourselves the following questions:

- When is phase 1 (ILM) finalized? This question became relevant because new initiatives placed new demands on the ILM. The enthusiasm of the development team is considerable, and one needs very little to pick up new requirements and include them in the technological plans. The problem is that AiREAS has neither resources nor funding of its own. The cooperation is purpose-driven through projects that produce measurable steps towards a "healthy city." Phase 1 was budgeted without those new issues. Any new proposition first needs to go through the three steps to get to financial backing. However, when a proposition is made for many people, it has already become part of an expected reality that they include in their talks. The ending of the ILM became postponed as phase 1 was continuously renewed with new requirements but without additional financial commitments. We needed to break through this impasse and determine the finalization of ILM phase 1, allowing for the start of ILM phase 2, or, alternatively, find indefinite funding. The latter was unlikely, even though we had proposed creating a start-up fund with government money as a type of loan.

This had not worked out yet. So the best way to deal with this was to close phase 1 properly, account for it, and open up workgroup discussions for the next phases.

- How do we interpret the data of the working ILM and deal with feedback and new requests coming from the new project ideas? Some parts of the ILM infrastructure need to remain fixed for medium term scientific research, while new scientific plans and feedback information suggest a remodeling of the network. Static versus dynamic becomes a point of potential friction. The installation of the ILM is could be considered a technological milestone, but for the scientists involved, it was only a starting point. They need a variety of data from multiple years to enable true interpretation for their research. On the other hand, the progressive nature of AiREAS towards a healthy city brings in new knowledge and views that demand the dynamic adjustment of the infrastructure. The handshake between the two extremes has fostered a continuous discussion.
- How to finance new projects? The Eindhoven city council and the Province invested in the ILM and its basic scientific research. Any new ideas needed to be funded themselves, and could not simply rely on the purse of the city. AiREAS had defined a royalty structure and also tried to link with innovative impulses that were generated by its open data, but these cyclic economies needed time to develop.

While all these issues were at hand, we started to look at the data provided by the ILM when it was released from Validation and Calibration in December 2013.

4.1.5   Interpreting the ILM Data

When the ILM became operational in September 2013, the only way to access the data was through an IP address from a database (see Chap. 3). This may be a valid procedure for professional users in the participating institutions, but for the general public, additional visualization was needed. Key in the initiation of AiREAS was the desire to involve citizens in their own healthy city development and the inherent responsibilities. AiREAS was, after all, a citizen's initiative. But how do you communicate in such a way that citizens react positively to innovation and their behavior? In the introduction to this manuscript, we have already referred to the levels of awareness of individual people and the lack of awareness of the masses. People today resonate to the hum of consumer- and money-based welfare. How can we open their minds to committing to the development of and contribution to their own wellbeing? How do we establish a society that opens up to change without fear of the unknown, strengthened through a sense of responsibility that starts with one's own perception of reality? Or can it be done in a different way? Making visible the invisible had opened up a whole array of research issues.

For instance: General, money-driven attitudes produce fears that if the pollution in a region becomes common knowledge, the prices of the houses might drop. This would justify avoiding openness in AiREAS communication. Other concerned people suggested that individuals with lung or heart problems would seek remuneration when scientific proof is made available about the effects of pollution on their health, especially when it becomes known that the State has reacted with reluctance in regard to their overall responsibility for pollution patterns while blindly focusing on economic growth. Comparisons were made with the tobacco industry and smoking, including the multibillion dollar claims against these enterprises.

4.1.6 The Transition

A new transition of governance became apparent. In the past, the prevailing social economic culture for welfare tended to avoid openness about certain environmental or humanitarian issues for fear of critics, financial drawbacks and economic growth impediments. The new tendency towards total openness and stimulating, self-regulating, wellbeing-based practice in the city can be seen as a breakthrough. This transformation in attitude was not supported everywhere, and there were certainly many who looked at the development with doubt and fear. But open data on the internet had already shown that cover-up strategies would never last and, in the end, would become a bigger political hazard than openness. After all, openness not only invites criticism but also encourages the taking of mutual responsibility for solutions, with all their innovative spin-offs. Eindhoven took the lead by accepting AiREAS as an instrument for change, but at the same time, needed to accept that it had to change itself too.

4.1.7 Communication

Being a multidisciplinary organization with civilian participation, the issue of communication became a serious new area of experimentation. We could distinguish three areas of attention right from day 1 of the operational working of the ILM:

1. Reading and interpretation of the near real time data
2. Communication with the city's population
3. Avenues of use for what we learned about the pollution patterns.

By the time we finalize this manuscript (June 2015), we will have 18 months of experience with these three points. It is an ongoing process that will get richer and richer as we proceed. As already stated, we need to define milestones. The ILM was designed to "make visible the invisible." At this stage, we can state safely that this mission is accomplished. The next step is to determine what to do with what we see that we couldn't see before. We must go step by step through these three points. This represents a powerful learning curve.

4.1.7.1 Reading and Interpreting the Near Real Time Data

The ILM had been designed by technology- and science-driven experts. You have already seen, at the end of Chap. 3, a list of new scientific research activities that were organized around the ILM. Technicians and researchers may have the knowledge to work with the raw data streams coming from an IP address, but the town's citizens, the AiREAS Sustainocrats and many others involved would need another human interface to visualize the air quality status. This was something we did not know yet and were about to find out. Such tools were not yet available for use.

ECN Tool

New Year and fireworks

The first event that triggered our curiosity was New Year's Eve, 2013/2014. ECN had the only self-made tools for looking at the real time and historical data. The first real time views of pollution of fireworks arose out of our enthusiasm for the potential of the ILM system. It was also the very first time that we got insight into the behavior of Ultrafine Particles (UFP) measured in 5 locations (Fig. 4.4).

The experience was tremendously positive, but instantly gave rise to the need for processed facilities and new information feeds for usage not just by ECN, but by AiREAS in general. The internal tooling of ECN was a first step but was not yet a tool for public use or for usage throughout AiREAS.

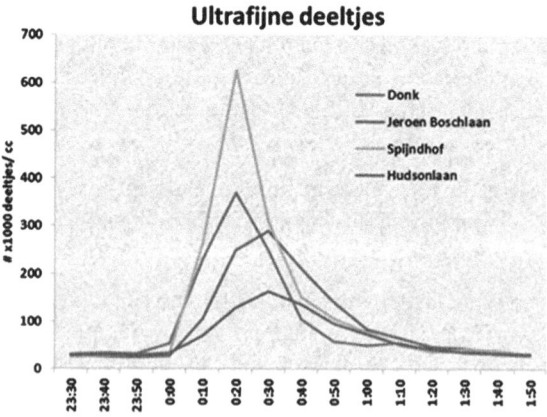

Fig. 4.4 New Year fireworks 2013/2014 UFP

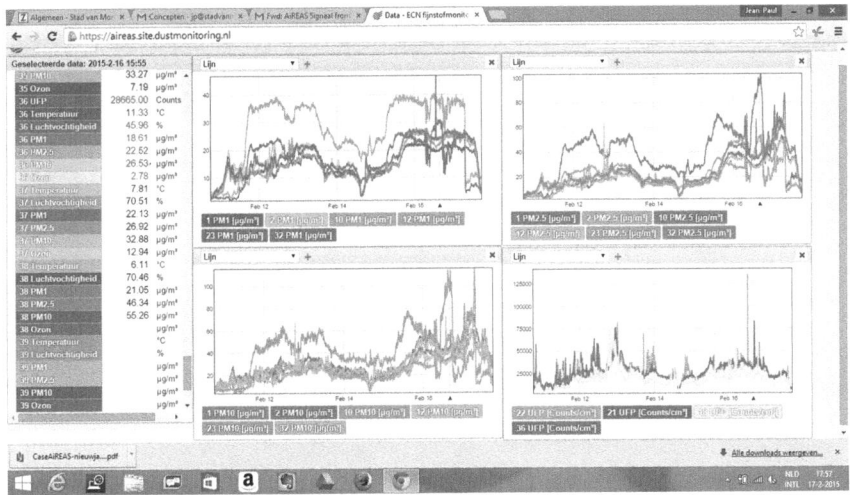

Fig. 4.5 A screenshot of the ECN tool

ECN suggested offering its internal tool for testing purposes to the AiREAS organization. The tool offered historical insight into the data and a dynamic graphical display option of data from each Airbox combined according to type of measured particle size or visualized in a mapped version (Fig. 4.5).

The tool was made available for limited use within the AiREAS partner and management structure to enable analysis of visualized data. This was indeed useful, especially when observations were required of high pollution suspects, such as the firework peaks. But the tool did not trigger curiosity or real time event monitoring.

The Imtech App

During the October 2013 meeting, Marco van Lochem successfully suggested developing an App for mobile phone usage. The normal routine would have been to go through the three steps of "the way things work at AiREAS". Then, the App would have received co-creation attention and a budget for development. To our surprise, Imtech had already taken on the challenge internally as a production of its own, and in June 2014, the App was presented as a teaser in a limited test edition (Fig. 4.6).

It was the App from Imtech that allowed for instant monitoring of the ILM network. The color code used by the App signified moments of intensified pollution by changing from green to orange and red. The border values for changing the colors were more or less in line with the norms used in the Netherlands, but by no means yet within an agreement. That was not possible because we had no idea yet what to agree to. The tool became a fundamental citizen's observatory that triggered curiosity in real time when colors changed. One simply had to develop the habit of

Fig. 4.6 A screenshot of the Imtech App (now Axians)

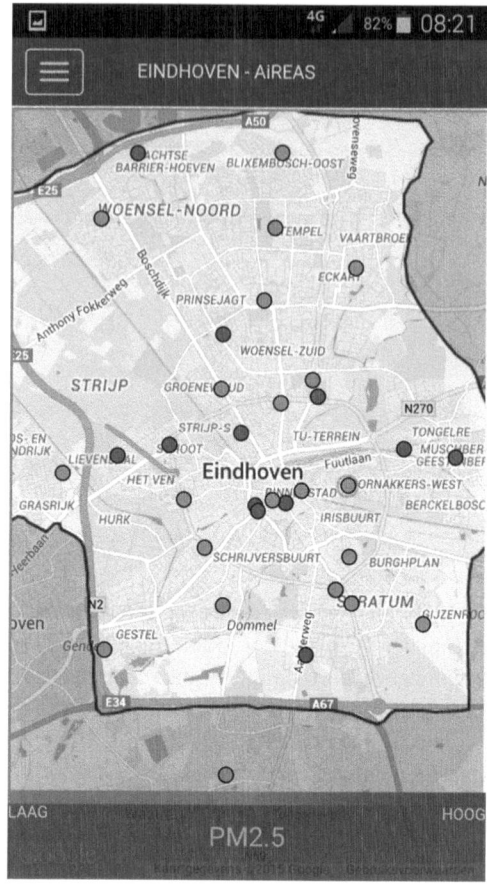

opening the App once in a while to check the status of the network. Then, someone who wanted to go into more depth of interpretation or analysis could take the ECN tool and work out some explanation.

The App initially worked with limitations and remained in the experimental phase. The first experiences were negative due to start-up installation problems and the incompleteness of the information supply. No pollution events had occurred yet that could possibly justify the App. A comparison was made with the air quality App of the RIVM (Ministry of Health). This App did not really contribute to awareness either nor did it trigger curiosity. Imtech claimed an investment of 70€ without AiREAS's coverage or perspective of reciprocity. It was frustrating for them that the effort had not gotten the enthusiastic backing that they felt it deserved. At the same time, it became clear that certain partners were still inclined to let their welfare mentality take over when they saw a chance. For most people involved, it became a learning process to distinguish between the two paradigms and make rational choices about when to apply one or the other.

Things changed drastically for the App when one summer morning, the first APP tester, co-founder Jean-Paul Close, looked at it and saw for the first time, to his surprise, a drastic change in coloring of the Airboxes. Nearly the entire city had turned orange and red. It was July 21, 2014. This called for further investigation. The entire team was notified about the phenomena and instant curiosity inspired people to analyse the situation in order to produce a preliminary interpretation and explanation for the curious event. A combination of massive BBQ'ing in the city, no wind, warm sunshine and high levels of humidity, had caused a peak of chemical reactions to occur in the air. These produced nausea, sickness and even death in certain people with lung problems.

This was the first time that an event in real time had been detected and opened as case for instant analysis with open feedback to the city via the blog, social media and the local news media. Without the App to trigger curiosity, this event could have passed unnoticed, ending up in the statistical averages without the possibility of cross-referencing it with real time environmental observations and the other key sensors available to us: our eyes, noses, ears, etc.

Instantly, the App gained status of key importance to AiREAS, to the satisfaction of Imtech, even though it had not yet been incorporated into a project with financial backing for development. The experience was positive, and discussion started on how to improve the App with the feedback and experiences obtained. This discussion, of course, may develop into and AiREAS project status, possibly with positioning on the EU scale with intended funding from H2020 unless Axians decides to keep the development and deployment to themselves.

The issue now arises that we probably would not have been able to develop as many insights as we have had Imtech (now Axians)[1] not made the decision to develop the App. The project-driven route would have eventually satisfied the financial backing of the early development, but this would have taken time in a setting in which no one knew what to expect. This time and awareness was gained thanks to the proactive attitude of Imtech/Axians, strengthening our ability for instant insights and the overall positioning of AiREAS in the early field of citizen's observatories. This too was an important lesson learned, and we trust that Axians will eventually be compensated by the effort.

Casus Collection

With the Imtech/Axians App, the ECN tool and the lively city equipped with the ILM, a whole series of observations were registered and documented during those 18 months. It was decided that every case would be described and shared among the teams for further elaboration. As of the writing of this document, the following cases have been registered:

---

[1]At the end of 2014, portions of the multinational Imtech were taken over by Vinci Energies in Paris and renamed Axians.

- *Firework peaks, 2014 and 2015 (two years, two different weather types)*

The culture of fireworks to celebrate the New Year has been shown to be highly polluting, especially in the ultrafine dust (UFP) spectrum.

- *Summer BBQ peak*

The combination of mass usage of barbecues with specific weather conditions displayed very surprising results and interpretations. The local hospital helped with the observation that similar situations had occurred in operation chambers when the burning technique of closing wounds would produce fumes that reacted with pollution from the street and high OZONE levels. It also produced nausea among the OC personnel.

- *Liberation day with 300 war vehicles (no peak)*

Every year, Eindhoven celebrates its liberation days. In 2014, the celebration had an extra dimension since it had been 70 years since the city had been freed by the allied forces. A huge festival was organized, with over 300 old time war vehicles. One would expect that such a massive parade of heavy trucks moving through the city would produce high levels of pollution. To our surprise, nothing of that was detected. Also, the noise levels of the engines of those trucks seemed to be much less than those in their modern counterparts. This suggests that the technology in World War II was much more sustainable than what followed it.

- *Light route (peak)*

In an additional celebration of the liberation, and as a tourist attraction, Eindhoven lives up to its name of the Light City by organizing the Route of Lights, a winding path of various illuminations throughout the city. This event lasts three weeks. The pollution peaks that we missed during Liberation Day were clearly visible during the Route of Lights. That part of town was highly polluted for the entire three weeks.

- *Torch event—Christmas peak*

Another popular event around Christmas is the Torch Light Parade. Thousands of citizens join together to carry torches along a specific route. This stands as a call for solidarity and social cohesion. This event was also clearly spotted in the measurements.

- *Different behavior UFP compared to PM > 1*

On various occasions, different behavior was detected between ultrafine particles and those of a larger size. UFP are generally produced by local events while anything larger tends to affect the entire city. The dispersion, particle behavior and reactions seem fundamentally different from the other type of particles.

- *Inversion—weather*

At a certain stage, high levels of pollution were detected for no apparent reason. Investigation led us to a weather phenomenon called "inversion". A cold air front presses on top of a hot layer below, compressing the air, including its pollution. The opposite occurs when the front has passed.

- *Possible strange situation Mauritsstraat (2 Airboxes nearby show totally different values)*

Why would two ILM stations located at a short distance from each other show fundamentally different values? Is this due to technical reasons or are local circumstances playing a role?

- *Possible agricultural cause of high peaks of pollution*

In March 2015, a sudden peak of pollution affected Europe entirely to the extent that big cities like Paris took remedial measures to close the city to certain traffic. No apparent cause could be detected until someone suggested that seasonal agricultural preparations of the land could be behind the peak.

These cases were collected and compared with similar situations over the years. Meanwhile, the cases have been presented to:

- the operational kernel of AiREAS to see what measures are possible to reduce the pollution
- the entrepreneurial community of Eindhoven to see if solutions can be found through technological innovations
- the scientific community to enhance our scientific insights and produce new projects for investigation and development of knowledge
- the public through open communication to trigger social innovation and awareness.

4.1.7.2 Communication with the Citizens

Let us jump back to December 2013. The positive decision to work out applications for mobile phones had not yet materialized into a project or a funding agreement. There was still discussion on what such an application should look like. Should we produce the end result as an APP or produce an API, an interface to which APPs could be related? The parallel decision by Imtech to produce an initial APP was an interesting case for seeing how such communication would work. But the App was not available yet by the end of 2013.

The only people who could monitor the network at this stage were ECN and those few partners that could deal with the direct data access link. There was, however, one place where the data was going to be displayed. This was the AiREAS website. The site development had been agreed upon as part of the first phase.

www.aireas.com

AiREAS Website

Who could imagine that the relative simplicity of setting up a website would become, in regards to communication, such a virtual tower of Babel? Right from the formal kick-off of AiREAS phase 1 (making the invisible visible) in October 2012, we tried to establish a communication team that could experiment with communication from a "persuasive" point of view. In the book "Sustainocracy, the new democracy,"[2] which describes Jean-Paul Close's process of awareness all the way up to the founding of AiREAS with Marco van Lochem and the formal kickoff at the city hall of Eindhoven in October 2012, the concept of "Burger-BAGE" is introduced. It is a concept for civilian involvement and alliance with the eco-system for "sustainable human progress," including health and air quality.

"Burger" means "civilian" and BAGE is an acronym of the following Dutch words, explained in English:

- **"B**ewustwording"—Awareness development
- **"A**anvaarding"—Acceptance of new responsibility
- **"G**edrag"—A change in behavior
- **"E**rkenning"—Reward for change of conduct.

The website was to experiment with these insights and produce awareness first. It would establish a relationship with the local citizens that would trigger the acceptance of responsibilities for the development of wellbeing. This type of dynamic in the website gave rise to many disputes and diversity in points of view. The building and maintenance of the website had not been budgeted for such complexity, and the people involved in the development were all small-scale entrepreneurs who could produce a simple website but refused to co-create the necessary communication skills through experimentation with new techniques and feedback. Since this type of awareness-driven *persuasive communication* is new, we could not find people with the sort of skills needed to be involved in the project. This meant that we had to develop the experience ourselves through trial and error.

Two elements of experience were crucial for the subsequent development of AiREAS:

1. *The website is an information tool, not a communication tool*

The website was recognized as a semi-static tool for supplying information, but not a system for communication. Communication requires human value-driven interaction between the sender and receiver, with feedback interpretation and experience development around the potential triggers of acceptance and societal change. A website is more of an online brochure in the world of welfare and trade. Social interaction is much more personal and a group process around wellbeing-based cohesion demands totally different tools and settings.

---

[2]Close (2012)—*Sustainocractie, de nieuwe democratie—MultiLibris.*

Fig. 4.7 The AiREAS website

With this insight in mind, the disputes were resolved, and after going through four different communication teams, we finally found rest and peace by placing the website's hosting and maintenance in the hands of a low cost support organization (Fig. 4.7).

The site provides information about AiREAS and shows a Google map of Eindhoven, with all the ILM measurement spots. Clicking on any spot displays the latest measurement data. Citizens can get insight in real time, but no historical information is provided (yet). Various citizens started using the information on the website by registering by hand every 10 min the relevant information for their own use. It was a start.

2. *Awareness is not the only factor for triggering action, as the majority of people are mere followers*

An AiREAS encounter in 2012 dedicated to civilian participation was hosted by the University of Technology at Eindhoven with the participation of Dr. Jaap Ham. Ham specializes in the psychological research of effects of technology on the behavior of human beings. When Jean-Paul Close explained Burger-BAGE, Dr. Ham stated immediately, to everyone's surprise: "No awareness! People are flock members, they follow the mainstream." This simple contribution had a major impact on the development of AiREAS. An example was used to sustain Ham's comment.

In the pursuit of energy transition, a lot of costly (welfare mentality) marketing was done to convince homeowners to install solar panels on their houses. There was no success until a young entrepreneur asked his house-owning uncle if he could place solar panels on his roof. The uncle agreed, and even convinced a neighbor to do the same. Within a few months, the entire street had solar panels.

With these two fundamental insights it had become clear that communication was to become an essential part of AiREAS. We needed to differentiate between the provision of static information via a website or press release, the marketing type of sales-oriented communication and the dynamics of communicating experiences, best practice and positive examples of behavioral change and innovation to trigger the population to follow. The early adapters ("me first") in every population may be acting from awareness and the desire to contribute to "a better world," but the mainstream population will only follow and produce a change of culture if they get acquainted with those changes, identify with the results by wanting to be a "me too"[3] part of it and can gain easy access.

The website issue had been solved and positioned, but now we needed to address the dynamics of true communication and its influencing potential. A new line of experimentation appeared, instantly influencing the preparations of the AiREAS POP (phase 2) and the way we communicated openly with and about our findings.

The Blog Became Our Dynamic Tool

Jean-Paul Close and the City of Tomorrow awareness programs had already developed a lot of experience in blogging.

- Jean-Paul's blog (in English): 5000 visitors average per year from 95 countries (mainly NL and USA)
- City of Tomorrow blog (in Dutch): 18,000 visitors average per year from over 40 countries, mainly NL, B, USA, Ge.

AiREAS activities had so far been shown and documented through the City of Tomorrow blog, along with all the other activities of the STIR Foundation. It was decided to set up a blog for AiREAS itself:

https://aireas.wordpress.com (Dutch)

While the blog was certainly open to anyone interested, it was written in Dutch specifically to address the community in Eindhoven. It was set up in 2013 and its number of visits has continued to grow, now averaging 4000 per year. The blog is directly linked to Twitter, LinkedIn and Facebook (Fig. 4.8).

An important effect of blogging and tweeting is the direct interaction with the local media, who pick up news items for processing in their written editions. Since the operational installation of the ILM and the direct, real time access to air quality information, we have received regular attention from the media when unique, interesting and curious insights have been discovered and shared through the blog.

---

[3]"Me1 (me first) and Me2 (me too)" positioning aspects as one of the 5 keys for success by Jean-Paul Close (guide for future market leaders—2005).

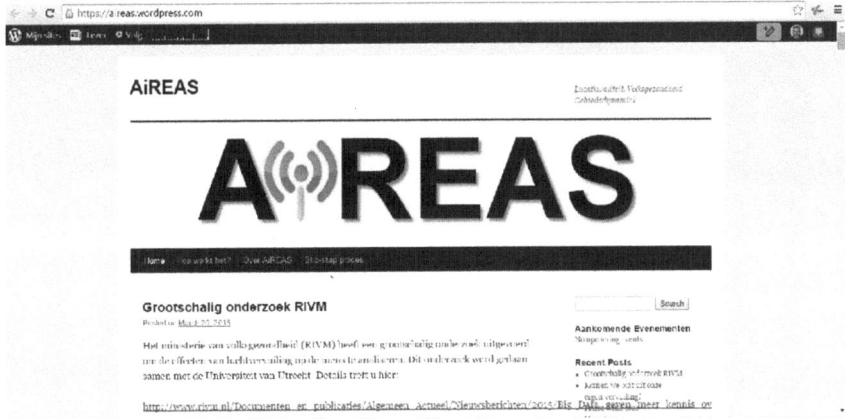

Fig. 4.8 The first AiREAS blog

Filming Progress via YouTube

The STIR Foundation was already in the habit of trying to visualize its initiatives through film. It was difficult enough to get participants into such simple processes as a congress, an encounter or a kick-off. Depicting our activities on film at least gave us the possibility of sharing the insights with a larger audience or integrating them into other forms of communication.

People don't tend to resonate with a new paradigm solely through words. Worldviews are simply too far apart. A video often explains much more and becomes a lasting document of a process. Recording critical events also helps as an educational tool when similar situations happen in new areas of attention. It provides people with a feeling of trust, as well as a sense of belonging, when they start to experiment in different ways with the new reality.

Here are some AiREAS recordings on YouTube:

The AiREAS concept in English: https://www.youtube.com/watch?v=iyxo6St YMw4

The same in Spanish: https://www.youtube.com/watch?v=LudujXawOCc

The APP demonstrated: https://www.youtube.com/watch?v=MBk5aVvj1wc

AiREAS real time data demo: https://www.youtube.com/watch?v=Jxig4 YxTF6w

Fireworks Impact 2014/2015 (In Dutch):
https://www.youtube.com/watch?v=PpvHHDGGR8Y

The local TV news item on the hanging of the first Airboxes: (In Dutch)
https://www.youtube.com/watch?v=2BwznuCGtwU

The complexity of getting to a commitment: (In Dutch)
https://www.youtube.com/watch?v=KxnuJz5y66E&list=PLBZBIkixHEidkOSv CbhJowtwJppiSU8OQ&index=1

Fig. 4.9 Students from Turkey help to explain the AiREAS concept to the Turkish community residents in Eindhoven

The formal announcement: (In Dutch) https://www.youtube.com/watch?v=sw8v Be-A9i0

The STIR HUB using AiREAS as an example: (In English): https://www.youtube.com/watch?v=cCPyOgc7S1I

Citizen Encounters

Using all the knowledge that we had gradually visualized, we organized numerous citizen encounters.

With Milieudefensie: A partner organization, Brabant Milieudefensie, is an NGO that also deals with air quality. We participated in each other's events to stimulate attention on air quality. More than 3000 signatures were collected in the interest of influencing city council elections.

With STIR Academy: This other STIR Foundation initiative organizes evening lectures and entrepreneurial encounters to stimulate the development of social and technological innovation. The STIR Academy experiments with the AiREAS coin, a value system that is given to local people who excel in their contribution to the field of health in the city. With it, they can follow the education programs of STIR Academy.

STIR Academy also became a European channel through the videoconferencing HUB platform and the Erasmus+ student exchange program (Fig. 4.9).

Through Business Partners: The "Dutch Leadership Trail"[4] visited AiREAS with a group of 20 CEOs. Axians organized the Internet of Things encounter on Eindhoven's High Tech Campus using AiREAS as a high tech example. AiREAS was selected as one of the potential finalists of the VINCI rewards. AiREAS has been invited to various encounters to speak about its views and method of working.

With FRE2SH: With this other STIR City of Tomorrow cooperative, dedicated to local quality productivity, tourism and self-sufficiency, new bicycle routes were developed. These routes connect points of interest to tell the story of sustainable

---

[4]The Dutch Leadership trail is an initiative organized by Camiel van Damme and Pierre Mellegers.

Fig. 4.10 The healthy city bicycle trail co-created with FRE2SH

human progress, varying from historical landmarks to good examples of entrepreneurship in value-driven processes, and even the site of an Airbox.

Air quality, innovation, co-creation, civilian participation and tourism based on health and quality of life have become instruments to link regions in the Netherlands and Europe through the Triple "i" (inspiration, innovation and implementation) platform of STIR Academy (Fig. 4.10).

This continuous interaction is slowly changing the way everyone looks at the city and its air quality. The tenor of the majority of the feedback we have received is one of worry. Many people feel helpless and don't really know how to address the issue. The question arises for AiREAS as to what we can do with the data and public/private commitment to make a difference.

4.1.7.3   What to Do with What We Learn About Pollution Patterns

When we look at the accumulated values around the public/private commitment to air quality and human health, we can already conclude the following:

- We can detect air pollution events in near real time and respond with observations to complete the casus. We can then use these specific cases to reflect and determine actions throughout the AiREAS partnership team.
- We can share this information with the public to stimulate:
  - social innovation
  - technological innovation

The data is shared among the AiREAS partners, which include the local government, scientists, business enterprises and civilians. Each may use the data for their own specific interests.

It has become clear that pollution is not just the byproduct of traffic and industry, but that many events and behavior-related issues in the city contribute as well. Scientists eventually may provide insight into what is healthy and what is not from an air quality and climate change point of view. Cultures have been built around lighting fires and burning things for human comfort and pleasure. Some of these issues can be overcome by introducing technological innovations, but many will require cultural modification around how we deal with our environment and our wellbeing. That is probably the most difficult issue to deal with.

Persuasive communication[5] has become a topic of discussion and an instrument to practice with. Persuasiveness is needed to achieve entrepreneurial backing through the development of innovations that make sense. These contributions will also use marketing channels as a way to help expand the movement. This type of communication has already become an area of scientific research: "How can we use technology to influence people?" In the public area of "safety in traffic," we already use technology extensively, but in all the other areas of key human wellbeing, as defined in Sustainocracy, we do not. Persuasion is sometimes perceived negatively, as it suggests "manipulation." When looking at our current society, manipulated as it is around capitalist hierarchies and dependencies, persuasion in the cause of awareness can be seen as a confrontation between interests: those who require blind submission and those who require aware participants.

36 % of the population in Eindhoven is worried about air pollution, but only 0.1 % actively takes action to do something about it. How do we increase that percentage of action? We have come up with various experimental trajectories that will be the subjects of new publications as they progress, representing new phases in our approach. They can be summarized into three key areas of attention:

1. Combining data from different sources, e.g., health, lifestyle, traffic, trees, weather and air quality
2. Further stimulating the innovation markets for new products and services, as well as social innovation patterns
3. Studying best practice in terms of persuasive communication techniques for mass involvement.

Spreading of AiREAS Values

The unique way of doing things at AiREAS and its higher purpose are recognized by all partners and made visible through publications, public presentations and representation in other cities. Breda was the second city to adopt the AiREAS

---

[5]Stiff, James Brian, and Paul A. Mongeau. *Persuasive communication*. Guilford press, 2003.

method by adding certain specific elements of its own, such as the effects of heat stress on the wellbeing of the city population.

The general partner assembly of AiREAS agreed in January 2014 that the cooperative had proven specific unique values to the world that can be expanded globally at this stage:

1. The multidisciplinary, sustainocratic, value-driven methodology of AiREAS
2. The experience and knowledge obtained in working with this method and the tension it sometimes produces with other paradigms
3. The ILM measurement system, with its important scientific contributions for modelling, data analysis and cooperative interpretation of multiple data feeds
4. The proof that wellbeing-based awareness generates new innovations and even business development for the welfare markets.

All this together is referred to as AiREAS phase 1. It can be adopted by other cities and regions as a self-contained package representing a lot of expertise and insight that no longer needs to be developed locally. With this basic phase 1 being readily available, any new city or region can concentrate on bringing in its own social, historical, cultural and demographic elements to produce authentic and unique spinoffs for the local community and market, as well as the international market.

4.1.8 Benchmarking and Referencing Our Practical Ideologies

While writing this analysis, we also began referencing our practical work with theories that had evolved elsewhere. We were already using many drawings in our text from the hand of co-founder Jean-Paul Close. Others have made drawing and models as well, and at this stage, it may be interesting to look at the contributions of Peter Senge[6] and Otto Scharmer[7] (best known for Theory U) who introduced the Ego to Eco matrix on the site of the Presencing Institute. It shows four levels of awareness that are similar to the Dabrowski layers of positive disintegration introduced in Chap. 1 (Table 1.1). The most interesting contribution of this matrix is in its presentation of this awareness at individual, group, institutional and global system levels (Fig. 4.11).

When we look at this matrix, we see the evolution of the AiREAS story all the way up to putting the level 4 'Awareness-based collective action' into practice. The key to Sustainocracy is that it can position any community-based society today

---

[6]Kofman, Fred, and Peter M. Senge. "Communities of commitment: The heart of learning organizations." *Organizational Dynamics* 22.2 (1993): 5–23.

[7]Scharmer, Claus Otto. "Theory U: Leading from the emerging future." *A Social Technology of Freedom (working title)* (2007).

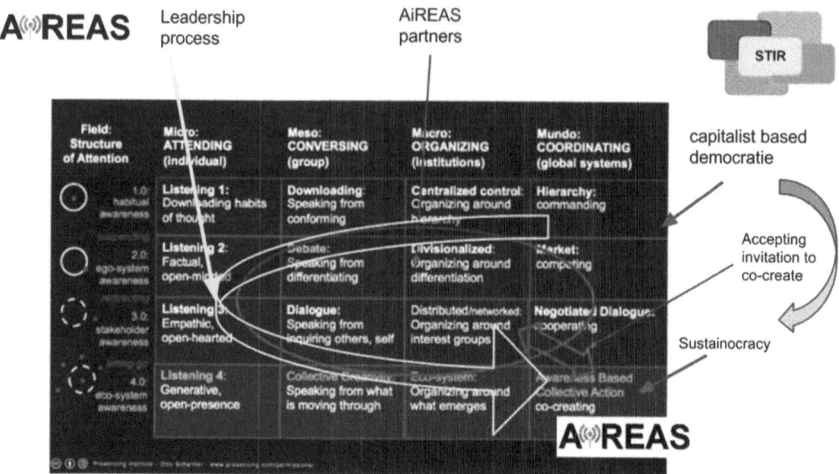

Fig. 4.11 The complexity and process of transformation (matrix is courtesy of the Presencing Institute, USA)

within that particular matrix quadrant. Community leadership is representative of commonly accepted key human values, as explained in Sustainocracy, through proven awareness and commitment. Individuals, institutions and entire community cultures may find themselves on any of the four levels of awareness. Those who emphasise eco-awareness are willing to consider participating in the co-creation platform of AiREAS. Some people have that level of personal empathy, but may work in institutions or cultures that do not. They either have enough authority to start down a transformation path with the institution, perhaps using participation in AiREAS as a guiding principle, or they don't have that authority and decide not to join the effort.

When comparing this with the publications of Dr. Kazimierz Dabrowski and his views on positive disintegration, we see that the layers are very similar to what the Presencing Institute uses in its matrix (Fig. 4.12).

It became clear that local platforms such as AiREAS depend very much on human beings who have reached the required level of awareness, combined with the level of professional authority to accept partnership in a coalition. Newcomers that lack the insight and awareness tend to disturb the process in the AiREAS setting until they break through or leave the group. Many people and institutions passed through AiREAS in this first phase, trying to connect from their level of awareness, but only those at level 3 and higher remained.

At the same time, we see the system's overall awareness develop within the old hierarchy, slowly letting go of the past while the AiREAS proof of principle began to prove itself as viable alternative. Meanwhile, a lot of longstanding impediments revealed themselves, typical of an ego- and competitive-led paradigm but useless and problematic for an eco-driven co-creative reality. Some of these impediments

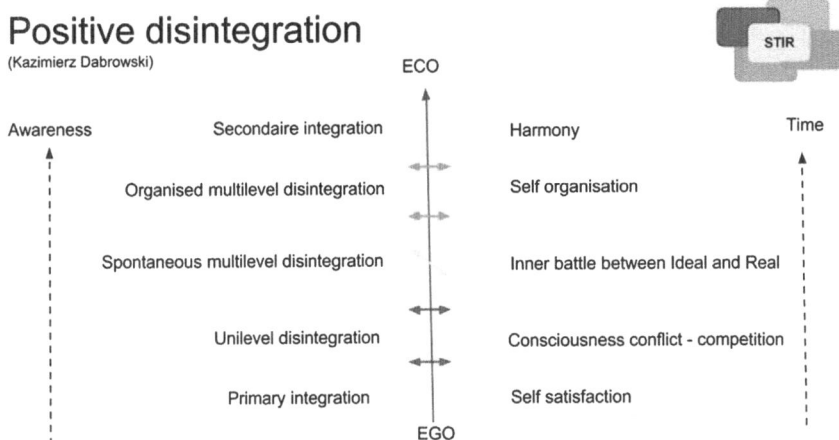

Fig. 4.12 These levels of awareness prove key to multidisciplinary co-creation

could be found in our constitutional development of rules and laws. The AiREAS process showed a lot of self-inflicted societal obstacles that needed to be addressed in order to make the new paradigm comfortably operational. Indeed, AiREAS was successful in making much more visible, often issues that previously had been considered normal and remained undisputed. Now, we could show a path forward, making clear the obstacles that needed to be removed. All the values that can be created need a level of freedom to come into existence. Connecting this at a later stage back into the world of transaction-based economics through royalties makes it worth the old system's while to facilitate the new paradigm and solve the obstacle issues.

4.1.9 The Royalty System

The human values-driven productivity of a Local AiREAS (city) provides the world with unique knowledge-based innovations that can be extended throughout the traditional commercial world. Since all innovations generated in an AiREAS cooperative contain the intellectual property of all participating members, a royalty is included in the global expansion of the values. The royalties are managed by Global AiREAS and revert back to the region of the Local AiREAS where the values were co-created. This way, the Local AiREAS is stimulated to keep innovating and calibrating its efforts to its own health and air quality development, as documented proof of principle for the world market.

The value creation dynamics and royalty scheme applied in AiREAS is referred to as the Pyramid Paradigm and was introduced into 21st century business development by Jean-Paul Close in 2007 (Fig. 4.13).

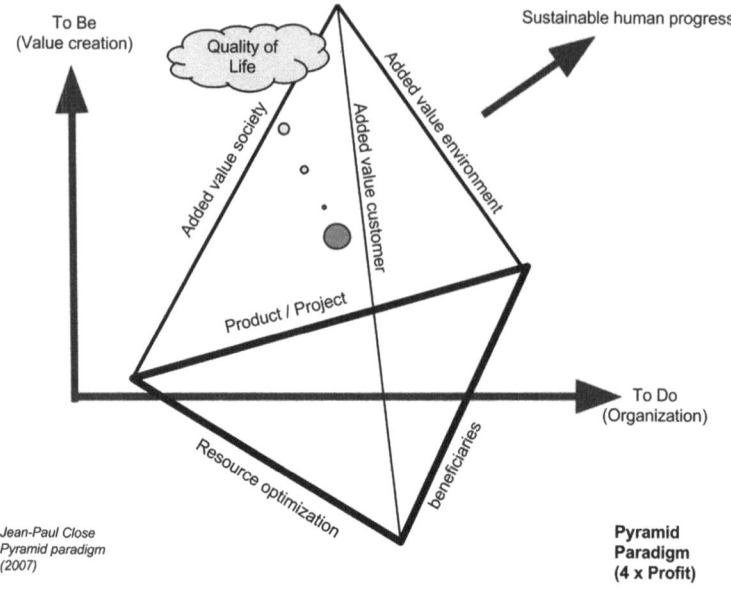

Fig. 4.13 The value-driven pyramid paradigm for business

4.1.10 Some of the Transformative Issues

During the evolution of AiREAS in Eindhoven, certain transformative issues were encountered. They proved that AiREAS was visualizing many more invisible things than just air pollution. Here, we list a few examples:

- **The cooperative that cannot be**: AiREAS is value-driven cooperation in which all partners engage for the higher purpose of co-creating healthy cities. A formal cooperative in Holland within the old paradigm can, however, only be legally set up to defend and grow the material interests of its members. All values co-created in AiREAS can, of course, be related back to economics once they have proven themselves, but this is a secondary consequence. The primary cooperative issue is the production of improved regional health and air quality. The secondary issue is connecting the values that were co-created to the trade system through the partners. It is highly debatable that a co-creation might be considered "illegal" by the reigning system. This demands a serious review of our systems of law and probably even our constitution, or at least the way we implement it in different paradigms.
- **Government cannot be a member**: The co-creation of a healthy city in a multidisciplinary setting cannot be envisaged without the participation of the city governance that controls tax collection, spending and infrastructure. But officially, the city government, in the present day role of constitutional executive, cannot be a member of the consortium if it is also co-financing it. In our

current legal setting of money-dependent hierarchies, the financial partner has a control function on the expenditure of the means. It cannot control itself. In AiREAS Eindhoven, an exception was made, but this, of course, needs to be formalized within our legal systems. As soon as a Local AiREAS produces its own innovation fund with the royalties obtained, then the problem disappears. The transformative issue is one that changes the government's position from that of a regionally dominant contractor into a facilitating partner for regional development.

- **Tax authority**: Initially, the tax authority did not accept AiREAS as a business enterprise that can reclaim value-added taxes. "You are an end user of what you purchase in the healthy city context and, hence, not liable to reclaim the taxes." Our arguments that our healthy city approach is a public/private Research and Development infrastructure for further expansion to the rest of the world, as well as a structure for triggering royalty-based entrepreneurial spin-offs, only landed when we produced our first invoices. Meanwhile, the tax authority had frozen the tax payback, causing a 21 % gap in the AiREAS phase 1 budget. It was eventually solved through Marco's administrative thoroughness and tenacity, but the transformative issue is needed to establish a solid case for eventual changes in the tax policies for genuine Sustainocratic functions. Recognizing the harmonization effects of the "transformation economy" as a basis for new economic growth impulses should do the trick. It is just a matter of time and further proof of concept for it to become part of the societal mainstream and its operational structures.
- **Business partners**: The business partners that committed to the AiREAS processes had to get used to the project and its result-driven way of working in the context of multidisciplinary wellbeing. AiREAS has no other initial means than those made available by the partners. Values can only be recognized and documented when they have proven themselves as measurable results within the "healthy city" context. AiREAS is hence not a customer, but a connecting instrument between means and goals to achieve a desired result. Planning and commitment are essential for building trust among the partners. If a commercially-oriented organization begins to ask for more money halfway through a project, then the overall partner relationship is upset and so is the execution of the project. Business partners had to get used to rewarding themselves through the global expansion potential of value creation, not trying to enrich themselves through the co-creation effort and the group's investment. This had a transformative effect. Some new age entrepreneurs came in directly from the value-driven perspective. Old-fashioned enterprises often disappeared or required adjustment to fit in.
- **Civilian partners**: Civilians in our city are used to dealing with a dominant, hierarchical government that makes decisions for the citizens. One is used to asking permission for everything related to governance. This permission-based culture of caretaking is broken when the citizens begin taking responsibility themselves, without the need to ask for permission. Permission-based systems are bureaucratic and base their decisions on the old rules of a money-driven

system. Responsibility within the common context of a "healthy city" needs no permission, since every initiative or innovation is a welcome contribution to the town.

- **Government follows civilian leadership**: When City of Tomorrow kicked off in 2009, we wanted to define "sustainability," resulting in our definition of "sustainable human progress," including civilian responsibilities. The 2009 Government in Eindhoven still managed the city from the perspectives of economic accessibility and growth, with innovation platforms directed at the world's mass markets, not self-usage or local proof of principle. In 2010, new elections allowed the city the chance to work on "sustainability, applied innovation and civilian participation." City of Tomorrow had already started the AiREAS "healthy city" project, to which the councilor personally connected and committed. In 2014, further new elections resulted in a city council prepared to commit to the "healthy city," while City of Tomorrow and AiREAS were already expanding worldwide. In Eindhoven, AiREAS had started to develop the eco-system of local self-sufficiency by combining City and Rural activities through FRE2SH. Gradually, local governance follows the calibration of activities based on new ethical values and insights for sustainable progress. For the individual, it is a choice; for a complex institution such as a city government, it is a transformative process based on a combination of defining policy leadership and working from civilian precedents.

- **Reward system for wellbeing**: Our financial systems are geared towards financial trade and welfare development. Wellbeing development is not rewarded. The question as to "why someone tightening screws in a factory is rewarded with money and a woman investing her time and effort in raising her children, the members of the next generation, is not?" resonated in AiREAS. People that have access to labor often have to commute between work and home using polluting mechanisms. And they get paid for it. Those who try to solve the issues through awareness-based co-creation are only compensated when products or services appear that help remediate the problem, not for their awareness-based social innovation or behavioral changes. It is difficult to understand the functioning of the reward system in a polluting context and the lack of it in a context of responsibility. We created an AiREAS coin as a catalyst for value creation. Its use could be compared with a farmer who puts in a lot of effort to make his crops grow successfully and harvests abundantly. If the farmer's efforts relate one to one to the harvest, then no additional value system is needed, the harvest is his/her value. When, however, the input is provided by hundreds of people who then have to wait some time before the harvest becomes available for sharing, a coin system helps as a reminder of the individual contribution as a key to sharing what has been achieved. The AiREAS coin does not compete with the Euro, it enhances it by developing values for enlargement and reciprocity for those who invested in it. Meanwhile, the coins could be invested locally in education, network encounters and local productivity-sharing, making the remaining coins even more valuable in their local circuit.

- **Lawful solidarity and ethics**: Citizens are lawfully obligated to sustain their local governance and system through the constitution. Laws have been adopted that demand people express their solidarity through tax and insurance systems. Ethical conflicts arise when the current system shows itself to be responsible for the pollution that it tries to remediate by taxing the same system's parameters, demanding more and more. The ethical conflict exists within the definition of ethics itself. Is it ethical to maintain lawful solidarity with an economic system that has proven itself to have reached a destructive level? Or do ethics mean commitment to ecological and anthropological harmony with our surroundings and the transformative challenge of adjusting the human system's dynamics through the evolution of awareness?
- **Wellbeing and welfare**: Gradually, the duality of a trade- and wellbeing-based system at a regional level could be balanced. Wellbeing would create innovations that welfare could expand worldwide. The transformation economy of change interacts proactively with the transaction economy of growth. By placing the emphasis on wellbeing-based change and not on growth, a new regional balance between productivity and consumption could be initiated.

4.1.11 Conclusion

AiREAS phase 1, making visible the invisible, opened eyes and awareness to much more than just air quality in the city. The entire transformative process of a city community that starts calibrating itself, its behavior and dynamics based on a new set of human and environmental values has become visible through the AiREAS Eindhoven process. Getting to this point has proven to be a warm, valuable and rewarding exercise for everyone. This sounds like an ending, but in reality, it is a milestone representing the beginning of everything. Having made the invisible visible in all its complexity and transformative dynamics, all parties may now find it much easier to interact towards further steps in the permanent healthy city objective. We have provided the living body of the city of Eindhoven with a nervous system and the very first real and artificial intelligence to work with it. This is just a start. Phase 2, the POP,[8] has started. And so have different working groups around key issues such as CalVal (Calibration and Validation in low cost, open access dynamics) and Persuasive Communications (how technology and awareness affects human behavior).

---

[8]POP (Proof of Principle) research and civilian participation project linking air pollution exposure to human health and lifestyle. This project started in January 2015 with 40 participants. It expects to optimize the complex processes involving many disciplines and up to 11 different databases for cross-referencing and persuasion, to expand it to 4000 citizens in Eindhoven and 4 Million in Europe.

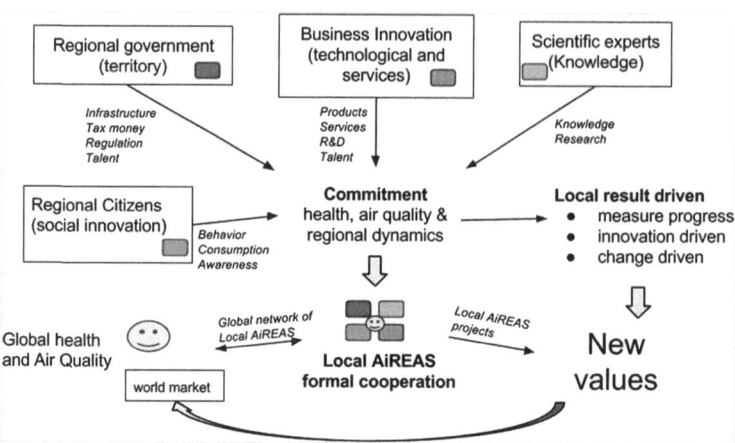

Fig. 4.14 Summarizing the new world of AiREAS

Phase 1 is providing enormous amounts of data for our universities to analyse and generate feedback. The entrepreneurial spinoffs are just a pioneering beginning of much more to come. Eindhoven will transform further and be an inspiring example for the entire world (Fig. 4.14).

I would like to close with a comment by our ICT database specialist and key phase 2 member John Schmeitz: "Everything we do in AiREAS is from the heart. We don't know the specific outcome up front but trust our venture and partners. All the values that appear ultimately reward us all in multiple ways, as they will all the generations to come."

Phase 1 has been successfully completed; phases 2 and 3 are on the way.

The AiREAS team (June 2015).

Erratum to: Potted Review of Economic Theory: The Complex Evolving System

Benjamin Aaron Rosen

Erratum to:
Chapter 1 in: J.-P. Close (ed.), *AiREAS: Sustainocracy for a Healthy City*, SpringerBriefs on Case Studies of Sustainable Development, DOI 10.1007/978-3-319-26940-5_1

The editor Mr. Jean-Paul Close would like to get his name removed from Chapter 1 of the book.

The updated original online version for this book can be found at 10.1007/978-3-319-26940-5_1

B.A. Rosen (✉)
University of Haifa, Haifa, Israel

© The Author(s) 2016   E1
J.-P. Close (ed.), *AiREAS: Sustainocracy for a Healthy City*,
SpringerBriefs on Case Studies of Sustainable Development,
DOI 10.1007/978-3-319-26940-5_5

Index

© The Author(s) 2016
J.-P. Close (ed.), *AiREAS: Sustainocracy for a Healthy City*,
SpringerBriefs on Case Studies of Sustainable Development,
DOI 10.1007/978-3-319-26940-5